LONDON MATHEMATICAL SOCIETY LECTURE NOTE SERIES

Managing Editor: Professor J.W.S. Cassels, Department of Pure Mathematics and Mathematical Statistics, University of Cambridge, 16 Mill Lane, Cambridge CB2 1SB, England

The books in the series listed below are available from booksellers, or, in case of difficulty, from Cambridge University Press.

34 Representation theory of Lie groups, M.F. ATIYAH *et al*
36 Homological group theory, C.T.C. WALL (ed)
39 Affine sets and affine groups, D.G. NORTHCOTT
46 p-adic analysis: a short course on recent work, N. KOBLITZ
49 Finite geometries and designs, P. CAMERON, J.W.P. HIRSCHFELD & D.R. HUGHES (eds)
50 Commutator calculus and groups of homotopy classes, H.J. BAUES
57 Techniques of geometric topology, R.A. FENN
59 Applicable differential geometry, M. CRAMPIN & F.A.E. PIRANI
66 Several complex variables and complex manifolds II, M.J. FIELD
69 Representation theory, I.M. GELFAND *et al*
74 Symmetric designs: an algebraic approach, E.S. LANDER
76 Spectral theory of linear differential operators and comparison algebras, H.O. CORDES
77 Isolated singular points on complete intersections, E.J.N. LOOIJENGA
79 Probability, statistics and analysis, J.F.C. KINGMAN & G.E.H. REUTER (eds)
80 Introduction to the representation theory of compact and locally compact groups, A. ROBERT
81 Skew fields, P.K. DRAXL
82 Surveys in combinatorics, E.K. LLOYD (ed)
83 Homogeneous structures on Riemannian manifolds, F. TRICERRI & L. VANHECKE
86 Topological topics, I.M. JAMES (ed)
87 Surveys in set theory, A.R.D. MATHIAS (ed)
88 FPF ring theory, C. FAITH & S. PAGE
89 An F-space sampler, N.J. KALTON, N.T. PECK & J.W. ROBERTS
90 Polytopes and symmetry, S.A. ROBERTSON
91 Classgroups of group rings, M.J. TAYLOR
92 Representation of rings over skew fields, A.H. SCHOFIELD
93 Aspects of topology, I.M. JAMES & E.H. KRONHEIMER (eds)
94 Representations of general linear groups, G.D. JAMES
95 Low-dimensional topology 1982, R.A. FENN (ed)
96 Diophantine equations over function fields, R.C. MASON
97 Varieties of constructive mathematics, D.S. BRIDGES & F. RICHMAN
98 Localization in Noetherian rings, A.V. JATEGAONKAR
99 Methods of differential geometry in algebraic topology, M. KAROUBI & C. LERUSTE
100 Stopping time techniques for analysts and probabilists, L. EGGHE
101 Groups and geometry, ROGER C. LYNDON
103 Surveys in combinatorics 1985, I. ANDERSON (ed)
104 Elliptic structures on 3-manifolds, C.B. THOMAS
105 A local spectral theory for closed operators, I. ERDELYI & WANG SHENGWANG
106 Syzygies, E.G. EVANS & P. GRIFFITH
107 Compactification of Siegel moduli schemes, C-L. CHAI
108 Some topics in graph theory, H.P. YAP
109 Diophantine analysis, J. LOXTON & A. VAN DER POORTEN (eds)
110 An introduction to surreal numbers, H. GONSHOR
111 Analytical and geometric aspects of hyperbolic space, D.B.A. EPSTEIN (ed)
113 Lectures on the asymptotic theory of ideals, D. REES
114 Lectures on Bochner-Riesz means, K.M. DAVIS & Y-C. CHANG
115 An introduction to independence for analysts, H.G. DALES & W.H. WOODIN
116 Representations of algebras, P.J. WEBB (ed)
117 Homotopy theory, E. REES & J.D.S. JONES (eds)
118 Skew linear groups, M. SHIRVANI & B. WEHRFRITZ
119 Triangulated categories in the representation theory of finite-dimensional algebras, D. HAPPEL
121 Proceedings of *Groups - St Andrews 1985*, E. ROBERTSON & C. CAMPBELL (eds)

122 Non-classical continuum mechanics, R.J. KNOPS & A.A. LACEY (eds)
124 Lie groupoids and Lie algebroids in differential geometry, K. MACKENZIE
125 Commutator theory for congruence modular varieties, R. FREESE & R. MCKENZIE
126 Van der Corput's method of exponential sums, S.W. GRAHAM & G. KOLESNIK
127 New directions in dynamical systems, T.J. BEDFORD & J.W. SWIFT (eds)
128 Descriptive set theory and the structure of sets of uniqueness, A.S. KECHRIS & A. LOUVEAU
129 The subgroup structure of the finite classical groups, P.B. KLEIDMAN & M.W.LIEBECK
130 Model theory and modules, M. PREST
131 Algebraic, extremal & metric combinatorics, M-M. DEZA, P. FRANKL & I.G. ROSENBERG (eds)
132 Whitehead groups of finite groups, ROBERT OLIVER
133 Linear algebraic monoids, MOHAN S. PUTCHA
134 Number theory and dynamical systems, M. DODSON & J. VICKERS (eds)
135 Operator algebras and applications, 1, D. EVANS & M. TAKESAKI (eds)
136 Operator algebras and applications, 2, D. EVANS & M. TAKESAKI (eds)
137 Analysis at Urbana, I, E. BERKSON, T. PECK, & J. UHL (eds)
138 Analysis at Urbana, II, E. BERKSON, T. PECK, & J. UHL (eds)
139 Advances in homotopy theory, S. SALAMON, B. STEER & W. SUTHERLAND (eds)
140 Geometric aspects of Banach spaces, E.M. PEINADOR and A. RODES (eds)
141 Surveys in combinatorics 1989, J. SIEMONS (ed)
142 The geometry of jet bundles, D.J. SAUNDERS
143 The ergodic theory of discrete groups, PETER J. NICHOLLS
144 Introduction to uniform spaces, I.M. JAMES
145 Homological questions in local algebra, JAN R. STROOKER
146 Cohen-Macaulay modules over Cohen-Macaulay rings, Y. YOSHINO
147 Continuous and discrete modules, S.H. MOHAMED & B.J. MÜLLER
148 Helices and vector bundles, A.N. RUDAKOV et al
149 Solitons, nonlinear evolution equations and inverse scattering, M.J. ABLOWITZ & P.A. CLARKSON
150 Geometry of low-dimensional manifolds 1, S. DONALDSON & C.B. THOMAS (eds)
151 Geometry of low-dimensional manifolds 2, S. DONALDSON & C.B. THOMAS (eds)
152 Oligomorphic permutation groups, P. CAMERON
153 L-functions and arithmetic, J. COATES & M.J. TAYLOR (eds)
154 Number theory and cryptography, J. LOXTON (ed)
155 Classification theories of polarized varieties, TAKAO FUJITA
156 Twistors in mathematics and physics, T.N. BAILEY & R.J. BASTON (eds)
157 Analytic pro-*p* groups, J.D. DIXON, M.P.F. DU SAUTOY, A. MANN & D. SEGAL
158 Geometry of Banach spaces, P.F.X. MÜLLER & W. SCHACHERMAYER (eds)
159 Groups St Andrews 1989 volume 1, C.M. CAMPBELL & E.F. ROBERTSON (eds)
160 Groups St Andrews 1989 volume 2, C.M. CAMPBELL & E.F. ROBERTSON (eds)
161 Lectures on block theory, BURKHARD KÜLSHAMMER
162 Harmonic analysis and representation theory for groups acting on homogeneous trees, A. FIGA-TALAMANCA & C. NEBBIA
163 Topics in varieties of group representations, S.M. VOVSI
164 Quasi-symmetric designs, M.S. SHRIKANDE & S.S. SANE
165 Groups, combinatorics & geometry, M.W. LIEBECK & J. SAXL (eds)
166 Surveys in combinatorics, 1991, A.D. KEEDWELL (ed)
167 Stochastic analysis, M.T. BARLOW & N.H. BINGHAM (eds)
168 Representations of algebras, H. TACHIKAWA & S. BRENNER (eds)
169 Boolean function complexity, M.S. PATERSON (ed)
170 Manifolds with singularities and the Adams-Novikov spectral sequence, B. BOTVINNIK
173 Discrete groups and geometry, W.J. HARVEY & C. MACLACHLAN (eds)
174 Lectures on mechanics, J.E. MARSDEN
175 Adams memorial symposium on algebraic topology 1, N. RAY & G. WALKER (eds)
176 Adams memorial symposium on algebraic topology 2, N. RAY & G. WALKER (eds)
177 Applications of categories in computer science, M.P. FOURMAN, P.T. JOHNSTONE, & A.M. PITTS (eds)
178 Lower K- and L-theory, A. RANICKI
179 Complex projective geometry, G. ELLINGSRUD, C. PESKINE, G. SACCHIERO & S.A. STRØMME (eds)
180 Lectures on ergodic theory and Pesin theory on compact manifolds, M. POLLICOTT

London Mathematical Society Lecture Note Series. 180

Lectures on ergodic theory and Pesin theory on compact manifolds

Mark Pollicott
Centro de Matemática
Universidade do Porto

CAMBRIDGE UNIVERSITY PRESS
Cambridge, New York, Melbourne, Madrid, Cape Town,
Singapore, São Paulo, Delhi, Tokyo, Mexico City

Cambridge University Press
The Edinburgh Building, Cambridge CB2 8RU, UK

Published in the United States of America by
Cambridge University Press, New York

www.cambridge.org
Information on this title: www.cambridge.org/9780521435932

First published 1993

A catalogue record for this publication is available from the British Library

Library of Congress Cataloguing in Publication data
Pollicott, Mark.
Lectures on ergodic theory and Pesin theory on compact manifolds
Mark Pollicott.
p. cm.
Includes bibliographical references.
ISBN 0–521–3593–5
1. Ergodic theory. 2. Manifolds (Mathematics) I. Title.
QA614.P65 1993
515′.42–dc20 92–32953 CIP

ISBN 978-0-521-43593-2 Paperback

Contents

Introduction 1

Part I. The basic theory

Chapter 1. Invariant measures and some ergodic theory 5

1.1 Invariant measures, 5
1.2 Poincaré recurrence, 9
1.3 Ergodic measures, 9
1.4 Ergodic decomposition, 10
1.5 The ergodic theorem, 12
1.6 Proof of the ergodic theorem, 15
1.7 Proof of the ergodic decomposition lemma, 18
Notes. 19

Chapter 2. Ergodic theory for manifolds and Liapunov exponents 21

2.1 The subadditive ergodic theorem, 21
2.2 The subadditive ergodic theorem and diffeomorphisms, 22
2.3 Oseledec-type theorems, 23
2.4 Some examples, 25
2.5 Proof of the Oseledec theorem, 31
2.6 Further refinements of the Oseledec theorem, 36
2.7 Proof of the subadditive ergodic theorem, 37
Notes. 40

Chapter 3. Entropy 43

3.1 Measure theoretic entropy, 43
3.2 Measure theoretic entropy and Liapunov exponents, 46
3.3 Topological entropy, 48
3.4 Topological entropy and Liapunov exponents, 50
3.5 Equivalent definitions of measure theoretic entropy, 53
3.6 Proof of the Pesin-Ruelle inequality, 58
3.7 Osceledec's theorem, topological entropy and Lie theory, 60
Notes. 62

Chapter 4. The Pesin set and its structure 63

4.1 The Pesin set, 64
4.2 The Pesin set and Liapunov exponents, 68
4.3 Liapunov metrics on the Pesin set, 69
4.4 Local distortion, 71
4.5 Proofs of Propositions 4.1 and 4.2, 73
4.6 Liapunov exponents with the same sign, 76
Notes. 77

An interlude 79

(a) Some topical examples, 79
(b) Uniformly hyperbolic diffeomorphisms:
(i) Shadowing, (ii) Closing lemma, (iii) Stable manifolds, 83
Notes. 85

Part II. The applications

Chapter 5. Closing lemmas and periodic points 87

5.1 Liapunov neighborhoods, 87
5.2 Shadowing lemma, 90
5.3 Uniqueness of the shadowing point, 94
5.4 Closing lemmas, 95
5.5 An application of the closing lemma, 96
Notes. 98

Chapter 6. Structure of "chaotic" diffeomorphisms 99

6.1 The distribution of periodic points, 99
6.2 The number of periodic points, 101
6.3 Homoclinic points, 103
6.4 Generalized Smale horse-shoes, 105
6.5 Entropy stability, 108
6.6 Entropy, volume growth and Yomdin's inequality, 110
6.7 Examples of discontinuity of entropy, 115
6.8 Proofs of propositions 6.1 and 6.2, 119
Notes. 122

Chapter 7. Stable manifolds and more measure theory 123

7.1 Stable and unstable manifolds, 123
7.2 Equality in the Pesin-Ruelle inequality, 127
7.3 Foliations and absolute continuity, 129
7.4 Ergodic components, 132
7.5 Proof of stable manifold theorem, 133
7.6 Ergodic components and absolute continuity, 137
Notes. 137

Appendix A. Some preliminary measure theory 139

Appendix B. Some preliminary differential geometry 145

Appendix C. Geodesic flows 151

References 155

Introduction

When I am dead, I hope it may be said
'His sins were scarlet, but his books were read'
(Hilaire Belloc)

The basic aim of this book is to present a simple and accessible account of some of the most basic ideas in the theory of non-uniformly hyperbolic diffeomorphisms, or more colloquially, 'Pesin theory'.

Part I consists of four chapters which contain basic material on the Oseledec theorem, the Ruelle-Pesin inequality, and the Pesin set. There is then a brief 'interlude' to mention some topical examples and to draw some motivation from the uniformly hyperbolic (or 'Axiom A') case. Then, Part II contains contains three chapters dealing with applications of this theory to periodic points, homoclinic points, and stable manifold theory.

In the course of the text I tried to bring out the following two themes

(i) *Generality*. Ultimately we want to arrive at a theory applicable to any smooth diffeomorphism of a compact surface (providing it has non-zero topological entropy);

(ii) *The rôle of measure theory*. In applying the theory it is remarkable how often invariant measures play a crucial role in situations where the hypothesis and conclusion are purely topological. In some sense, the Poincaré recurrence of invariant measures seems to compensate for the absence of the compactness often take for granted in uniformly hyperbolic systems.

This text is based on a short series of lectures I gave in the *Centro de Matematica do INIC na Universidade do Porto* between March and June 1989. These lectures were intended to give a basic introduction to some of the simpler and more accessible aspects of the theory (both for the benefit of the audience and myself). My choice of presentation was chiefly influenced by the more topologically oriented approaches in the work of Anatole Katok and Sheldon Newhouse. Because of the nature of the audience I was able to assume a solid background in undergraduate analysis, but not necessarily any specialist knowledge of either measure theory or ergodic theory.

In order to keep the exposition as breezy as possible I have unashamedly resorted to using two devices: (a) postponing the more tedious proofs to the ends of the relevant chapters (or indefinitely); and (b) restricting proofs to the case of surfaces if it provides a significant saving in effort.

Existing literature. There exist a number of very good accounts of certain parts of the theory, although these tend to be somewhat scattered in the literature, and for some topics the original articles still provided the best sources. At present the only textbook account of Pesin theory known to me is contained in the last chapter of *Teoria Ergodica*, by Ricardo Mañé [**Mañé$_2$**] (This book now has an English translation, but my references are to the original Portuguese language edition). Along with the preliminary sections of research articles by Katok [**Katok**] and Newhouse [**Newhouse$_{1,2}$**] this probably gives the clearest introduction to the foundations of the subject.

In addition, there exists a (currently) unfinished monograph by A. Katok and L. Mendoza which is already becoming a standard reference and promises to be a very useful introduction to the subject [**Kat-Men**]. I should also mention that there is a very comprehensive book by Katok and Strelycn [**Kat-Str**], but this goes far beyond the scope of an expository account.

These sources are nicely complemented by a clear account of the basic theory of stable manifolds contained in the survey article of Fathi, Herman and Yoccoz [**Fa-He-Yo**] (although in these notes I follow the last part of Mañé's article [**Mañé$_1$**] in order to preserve a slightly more topological flavor). In the last chapters we lightly touch upon the important, but harder, topics of absolute continuity, continuity of entropy and the C^∞ entropy conjecture. For these matters there now exist good accounts by Pugh-Shub [**Pug-Shu**] and Gromov [**Gromov**].

Finally, an interesting over-view of the theory (without any proofs) can also be found in a survey of Pesin and Sinai [**Pes-Sin**].

For the background material on ergodic theory there are a number of excellent modern references, of which my favorites are *An Introduction to Ergodic Theory* by Peter Walters [**Walters**] and *Topics in Ergodic Theory* by William Parry [**Parry**].

Acknowledgements. I would like to thank all the members of the audience of the original course, from which these notes are taken, for their very active participation. I am also very grateful to the anonymous referees for their very useful comments, to Sr[a] D. Zulmira Couto for typing the first draft from very rough notes, and to Luisa Magalhães and Viviane Baladi for their brave attempts to help me to reduce the number of errors and inconsistencies in the text. I would appreciate being notified of any further errors that readers may notice.

I am also grateful to David Ruelle for some encouraging comments on a preliminary version of this book, and to the authors of [**Kat-Men**] for making their notes available prior to publication.

Finally, at all stages of this work I was an investigador of the Instituto Nacional de Investigação Cientifica (INIC), an institution for which I have great respect and affection, and to which I would like to dedicate this book.

Chapter 1

Invariant measures and some ergodic theory

In this first chapter we shall describe some basic ideas in the ergodic theory of general measurable spaces. In later sections we shall specialize to diffeomorphisms of manifolds.

For the initiated we have included a more recent proof of the ergodic theorem, to help relieve the tedium. At the other extreme we have added an appendix (Appendix A) explaining some of the necessary background in measure theory.

1.1 Invariant measures.

In measure theory the basic objects are *measurable spaces* $(X,\mathcal{B})$, where X is a set and $\mathcal{B}$ is a sigma algebra (or σ-algebra). In ergodic theory the basic objects are measurable spaces $(X,\mathcal{B})$ *and* a measurable transformation $T: X\to X$ (i.e. with $T^{-1}\mathcal{B}\subset\mathcal{B}$). Given such a transformation $T: X\to X$ we want to consider those probability measures $m: \mathcal{B}\to\mathbb{R}^+$ which are 'appropriate' for T in the following sense.

Definition. A probability measure m is called *invariant* (or more informatively, T-*invariant*) if $m(T^{-1}B) = m(B)$ for all sets $B\in\mathcal{B}$.

This definition just tells us that the sets B and T^{-1}B always have the same 'size' relative to the invariant measure m. It is easy to see that if the transformation T is a bijection and its inverse $T^{-1}: X\to X$ is again measurable then the above definition is equivalent to asking that $m(B) = m(TB)$ for all sets $B\in\mathcal{B}$.

An alternative formulation of this definition is to ask that $T^*m=m$ where $T^*:\mathcal{M}\to\mathcal{M}$ is the map on the set $\mathcal{M}$ of all probability measures on X defined by $(T^*m)(B) = m(T^{-1}B)$, for all $B\in\mathcal{B}$.

Notation. We shall denote the set of all T-invariant probability measures on X by $\mathcal{M}_{\text{inv}}$.

We shall now give a trivial lemma which just gives another reformulation of the definition.

Lemma 1.1 (Characterizing invariant measures). The following statements are equivalent:

(a) $m \in \mathcal{M}_{\text{inv}}$;

(b) $\int f \circ T \, dm = \int f \, dm$, for all $f \in L^1(X, \mathcal{B}, m)$.

Proof. (b)$\Rightarrow$(a). For any set $B \in \mathcal{B}$ let $f = \chi_B$ be the characteristic function defined by

$$\chi_B(x) = \begin{cases} 1 \text{ if } x \in B \\ 0 \ \text{ otherwise} \end{cases}$$

We then have that $m(B) = \int f \, dm = \int f \circ T \, dm = m(T^{-1}B)$, and by definition $m \in \mathcal{M}_{\text{inv}}$.

(a)$\Rightarrow$(b). Starting with a measure $m \in \mathcal{M}_{\text{inv}}$ we have that $\int \chi_B \circ T \, dm = m(B) = m(T^{-1}B) = \int \chi_B \, dm$, for any set $B \in \mathcal{B}$. Thus by approximating any function $f \in L^1(X, \mathcal{B}, m)$ (in the L^1 norm) by finite combinations of these characteristic functions the result follows. □

We shall now begin to build up a collection of stock examples of transformations and invariant measures.

Examples. (i) Let $X = [0,1)$ be the half open unit interval with the usual Borel σ-algebra. For our transformation we choose $T: X \to X$ to be the fractional part of x multiplied by ten: $Tx = 10x$ (modulo unity). This transformation has many invariant measures, from which we shall choose m to be the usual Lebesgue measure.

To see that m is invariant we first consider intervals of the form $[a,b)$ for which we have that $T^{-1}[a,b) = \bigcup_{i=0}^{9} [\frac{a+i}{10}, \frac{b+i}{10})$ and therefore

$$m(T^{-1}[a,b)) = \sum_{i=0}^{9} \left| \frac{b+i}{10} - \frac{a+i}{10} \right| = b - a = m([a,b)).$$

Once we have this result for intervals (and the algebra of all their finite unions) we can deduce the same for the the whole σ-algebra $\mathcal{B}$ by wheeling out the machinery of the extension theorem (which we describe in Appendix A). A similar approach can be applied to all of the following examples.

(ii) Let $X=(0,1)$ be the open unit interval with the usual Borel σ-algebra. We can choose our transformation $T: X \to X$ to be the *Gauss map* defined by $Tx = 1/x$ (modulo unity). An interesting choice of invariant measure is the Gauss measure defined by

$$m(B)=\int_B \frac{1}{\log 2\,(1+x)}\,dx,\ B \in \mathcal{B}.$$

Here m is equivalent to the usual Lebesgue measure. (NB. To make T well defined we *should* throw out points $\{1/n \mid n>0\}$, etc. But since these are a set of zero measure (relative to m) we shall just adopt the convention of ignoring them.)

(iii) Let $X=\mathbb{R}^2/\mathbb{Z}^2$ be the standard flat torus and let $\mathcal{B}$ be the usual Borel sigma algebra. We define the transformation $T: X \to X$ to be a rotation in each of the co-ordinates of the form $T(x_1,x_2)=(x_1+\alpha_1,x_2+\alpha_2)$ where $\alpha_1,\alpha_2 \in \mathbb{R}$ (Figure 1). The usual Lebesgue-Haar measure on X is invariant.

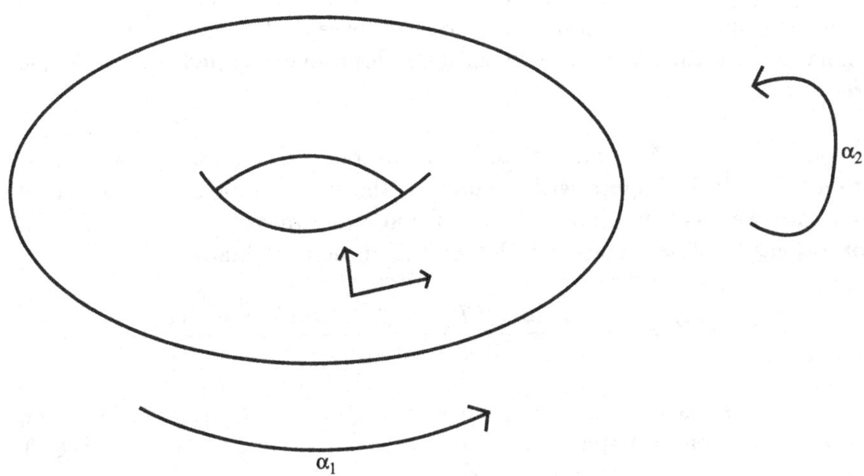

Figure 1: Irrational rotations on tori

(iv) Finally, we have the following trivial examples:

(a) Let $(X,\mathcal{B})$ be any measurable space and let $T: X \to X$ be the identity transformation. Therefore, *all* probability measures are invariant and $\mathcal{M} = \mathcal{M}_{inv}$;

(b) Let X be any set and let $\mathcal{B} = \{\emptyset, X\}$ be the trivial sigma algebra. There exists exactly one probability measure m defined by $m(X) = 1$, $m(\emptyset) = 0$ which will be invariant for any measurable transformation $T: X \to X$. Therefore, $\mathcal{M} = \mathcal{M}_{\text{inv}} = \{m\}$.

Standard convention. Whenever we state a result for *almost all* $x \in X$, with respect to the probability measure m (usually abbreviated to $a.a.(m)$ $x \in X$) it means that there exists $\Omega \in \mathcal{B}$ with $m(\Omega) = 1$ for which the result holds for $x \in \Omega$. In most of our statements involving measures, this condition is implicit (even when not explicitly stated).

We now come to a rather basic question:

When does a transformation $T: X \to X$ actually have an invariant measure?

Fortunately, in all of the cases we will be interested in this turns out not to be a problem at all, by virtue of the following lemma (whose proof is both short *and* sweet).

Lemma 1.2. (Existence of invariant measures). If $T: X \to X$ is a homeomorphism of a compact topological space and $\mathcal{B}$ is the usual Borel sigma algebra then there always exists at least one invariant measure (i.e. $\mathcal{M}_{\text{inv}} \neq \emptyset$).

Proof. The space $\mathcal{M}$ of all probability measures is a convex, compact, non-empty, topological space with respect to the weak* topology (a beautiful fact that we recall in Appendix A). Choose any measure $m \in \mathcal{M}$ and then for any $n \geq 1$ define a measure $m^{(n)}$ by the affine combination

$$m^{(n)} = \frac{m + T^* m + (T^*)^2 m + \ldots + (T^*)^{(n-1)} m}{n}$$

Since $\mathcal{M}$ is a convex space we see that $m^{(n)} \in \mathcal{M}$. Furthermore, since $\mathcal{M}$ is a compact space there must exist an accumulation point $\widetilde{m} \in \mathcal{M}$ for the sequence $\{m^{(n)}\}_{n=1}^{+\infty}$ i.e. $\widetilde{m} = \lim_{i \to +\infty} m^{(n_i)}$, for some subsequence $m^{(n_i)}$, $i \geq 1$. Finally, we can conclude that $\widetilde{m} \in \mathcal{M}_{\text{inv}}$ (and therefore $\mathcal{M}_{\text{inv}} \neq \emptyset$) since we can write

$$T^* \widetilde{m} = \lim_{i \to +\infty} T^* m^{(n_i)} = \lim_{i \to +\infty} \left(m^{(n_i)} + \frac{m - (T^*)^{n_i} m}{n_i} \right) = \lim_{i \to +\infty} m^{(n_i)} = \widetilde{m}$$

(NB. This proof is just a simple version of the Schauder fixed point theorem.) □

1.2 Poincaré Recurrence.

One of the most basic results, and probably also one of the most useful, in ergodic theory is the following.

Theorem 1.1 (Poincaré recurrence). Let $T: X \to X$ be a transformation on a measurable space $(X, \mathfrak{B})$ and let $m \in \mathcal{M}_{\text{inv}}$ be an invariant measure. For any set $B \in \mathfrak{B}$ almost all points $x \in B$ return to B under some iterate of T (i.e. $m(F)=0$ where $F = \{b \in B \mid T^n b \notin B, \forall n \geq 1\}$).

Proof. By definition, we have that $T^n F \cap B = \emptyset$ $\forall n \geq 1$ and so since $F \subset B$ it trivially follows that $T^n F \cap F = \emptyset$ for $n \geq 1$. By repeatedly applying the inverse of T we get that $T^{-(n-k)} F \cap T^{-k} \subseteq T^{-k}(T^n F \cap F) = \emptyset$, for $n \geq k \geq 0$ and we can deduce that the sets $\{T^{-k} F\}_{k=0}^{+\infty}$ are all pairwise disjoint. Finally, we can write that

$$1 \geq m\left(\bigcup_{n=0}^{+\infty} T^{-n} F\right) = \sum_{n=0}^{+\infty} m(T^{-n} F) = \sum_{n=0}^{+\infty} m(F),$$

since by T invariance of the measure m we have $m(T^{-n} F) = m(F)$ for $n \geq 1$. Thus, we deduce that $m(F)=0$. □

1.3 Ergodic measures.

Now that we have mastered invariant measures we shall turn our attention to a particularly important type of invariant measure.

Definition. An invariant measure $m \in \mathcal{M}_{\text{inv}}$ is called *ergodic* if whenever $T^{-1} B = B$, for some $B \in \mathfrak{B}$, then either $m(B)=0$ or $m(B)=1$.

This definition just says that the only set B which also equals $T^{-1} B$ must have the same 'size' as the whole space X or be trivial, *relative to the invariant measure m*. It is easy to see that if the transformation T is a bijection and its inverse $T^{-1}: X \to X$ is again measurable then the above definition is equivalent to saying that whenever $B = TB$, for some set $B \in \mathfrak{B}$, then $m(B) = 0$ or $m(B) = 1$.

Notation. We shall denote the set of ergodic measures by $\mathcal{M}_{erg} = \{m \in \mathcal{M}_{inv} \mid m \text{ is ergodic}\}$.

We shall now give a trivial lemma which reformulates this definition.

Lemma 1.3 (Characterizing ergodic measures). The following statements are equivalent:
(a) $m \in \mathcal{M}_{erg}$;
(b) $f \circ T = f$, for some $f \in L^1(X, \mathcal{B}, m) \Rightarrow f$ is constant a.e.(m).

Proof. (b)$\Rightarrow$(a). Given a set $B \in \mathcal{B}$ satisfying $T^{-1}B = B$ we define $f = \chi_B$ to be the characteristic function of B. The condition on our set B implies that $f \circ T = f$ and so we deduce that f is constant (up to a set of measure zero). For a characteristic function this means the value is either 0 or 1, i.e. $m(B)=0$ or $m(B)=1$.

(a)$\Rightarrow$(b). If we could find a function $f \in L^1(X, \mathcal{B}, m)$ which was *not* constant (up to a set of measure zero) then we could choose a real number c such that the set $B = \{x \in X \mid f(x) \geq c\}$ has measure $0 < m(B) < 1$. However the assumption $f \circ T = f$ implies that $T^{-1}B = B$, which leads us into a contradiction. □

A very reasonable question to ask at this stage is:

What makes ergodic measures particularly interesting?

We want to give two different answers to this question in the next two sections.

1.4 Ergodic decomposition.

Problems about arbitrary invariant measures in $\mathcal{M}_{inv}$ can frequently be reduced to (simpler) problems about ergodic measures in $\mathcal{M}_{erg}$ (at least when X is a topological space).

Given any compact, convex topological space C we define the ***extremal points*** to be the set

$$\text{Ext}(C) = \{x \in C \mid x = \alpha x_1 + (1-\alpha)x_2,\ 0 \leq \alpha \leq 1,\ x_1 \neq x_2 \Rightarrow \alpha = 0 \text{ or } \alpha = 1\}.$$

It is easy to see that if $C \neq \emptyset$ then $\text{Ext}(C) \neq \emptyset$ (Figure 2).

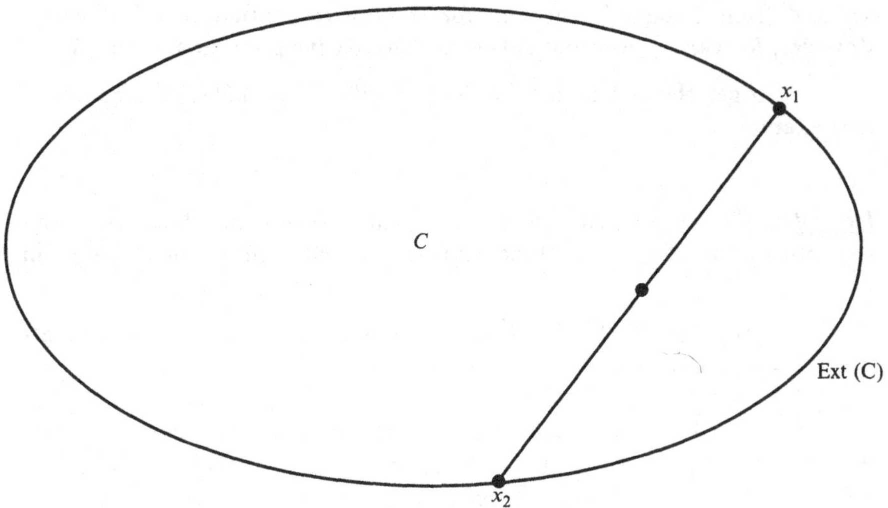

Figure 2: Convex sets and extremal points

With the choice $C = \mathcal{M}_{\text{inv}}$ we have the following useful result.

Lemma 1.4 (Ergodic decomposition). The extremal points of $\mathcal{M}_{\text{inv}}$ are precisely $\mathcal{M}_{\text{erg}}$ (i.e. $\text{Ext}(\mathcal{M}_{\text{inv}})=\mathcal{M}_{\text{erg}}$). Furthermore, given any invariant measure $m \in \mathcal{M}_{\text{inv}}$ there exists a probability measure μ_m on the space $\mathcal{M}_{\text{inv}}$ such that:

(a) $\mu_m(\mathcal{M}_{\text{erg}}) = 1$; and

(b) $\int_X f \, dm = \int_{m' \in \mathcal{M}_{\text{erg}}} \left(\int_X f \, dm' \right) d\mu_m(m')$, for any $f \in L^1(X, \mathcal{B}, m)$.

(i.e. the invariant measure m is an affine combination, weighted by μ_m, of *ergodic* measures m').

The first part of this lemma is an easy exercise (based on the observation that distinct ergodic measures are mutually singular, see

Appendix A for definitions). The second part of the lemma is usually deduced from Choquet's theorem for the abstract situation (cf.[Phelps]). However, for variety we shall outline an alternative proof in section 1.7.

To get the feel of ergodic decomposition we shall give a couple of easy examples.

Examples. (i) If we start from an ergodic measure $m \in \mathcal{M}_{erg}$ then the measure μ_m on $\mathcal{M}_{inv}$ is a Dirac measure concentrated on the single point m.

(ii) Let $T: \mathbb{R}^2/\mathbb{Z}^2 \to \mathbb{R}^2/\mathbb{Z}^2$ be the rotation in each co-ordinate $T(x_1, x_2) = (x_1 + \alpha_1, x_2 + \alpha_2)$ we mentioned before. The values we choose for $\alpha_1, \alpha_2 \in \mathbb{R}$ have a radical effect on $\mathcal{M}_{inv}$ and $\mathcal{M}_{erg}$. If the ratio $\frac{\alpha_1}{\alpha_2}$ is irrational then the usual Lebesgue-Haar measure μ is the unique invariant measure i.e. $\mathcal{M}_{inv} = \mathcal{M}_{erg} = \{\mu\}$. If $\frac{\alpha_1}{\alpha_2}$ is rational then $\mathcal{M}_{inv}$ and $\mathcal{M}_{erg}$ are both infinite and $\mathcal{M}_{inv} \neq \mathcal{M}_{erg}$.

We saw before that homeomorphisms of compact topological spaces always have invariant measures i.e. $\mathcal{M}_{inv} \neq \emptyset$. Since convex sets always have at least one extremal point we get the following result for free from the previous lemma:

Lemma 1.5 (Existence of ergodic measures). If $T: X \to X$ is a homeomorphism of a compact topological space and $\mathcal{B}$ is the usual Borel σ-algebra then there always exists at least one ergodic measure (i.e. $\mathcal{M}_{erg} \neq \emptyset$).

1.5 The ergodic theorem.

Ergodic measures have very good recurrence properties, which allow us to sharpen the result of the Poincaré recurrence theorem to say that *most* orbits are evenly distributed in the set X. Thus we come to the second important result in ergodic theory.

Theorem 1.2 (Birkhoff Ergodic Theorem). Let $T: X \to X$ be a transformation on a measurable space $(X,\mathcal{B})$ and let $m \in \mathcal{M}_{erg}$ be an ergodic measure. For almost all points $x \in X$ (with respect to the ergodic measure m) we have that

$$\frac{1}{n}\sum_{i=0}^{n-1} f(T^i x) \longrightarrow \int f\,dm, \quad \text{as } n \longrightarrow +\infty,$$

for any $f \in L^1(X,\mathcal{B},m)$.

If we choose a set $B \in \mathcal{B}$ with $m(B)>0$ and let $f=\chi_B$ be the associated characteristic function then we get the following interpretation of the theorem (cf. Figure 3).

Corollary 1.2.1 Given a set $B \in \mathcal{B}$ with $m(B)>0$, we have that for almost all points $x \in X$ (with respect to the ergodic measure m) the proportion of the orbit spent in B is given by $m(B)$, i.e.

$$\frac{\#\{\, 0 \leq i \leq n-1 \mid T^i x \in B\}}{n} \longrightarrow m(B), \quad \text{as } n \longrightarrow +\infty.$$

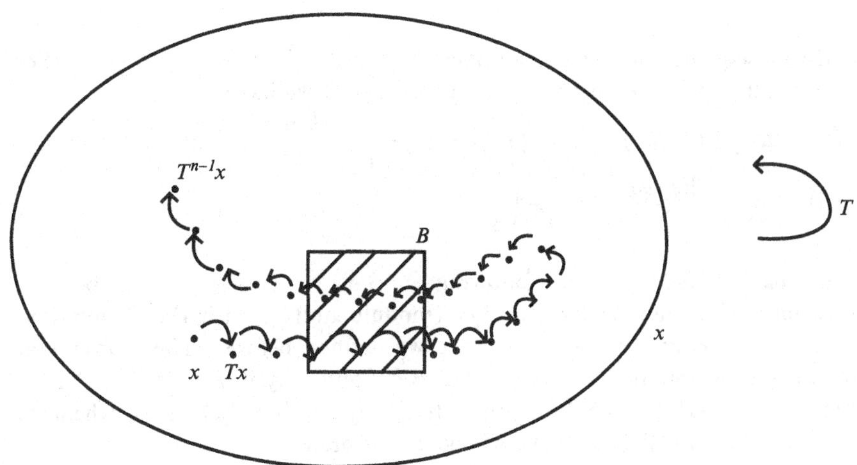

Figure 3: Proportion of orbit spent in B

Before giving the proof of Theorem 1.2, we mention some standard applications.

Applications. (a) For almost all $0 \leq x \leq 1$ (with respect to Lebesgue measure) the average number of zeros in the decimal expansion $x = 0.x_1 x_2 x_3 \cdots$ is 1/10 (such numbers are called *normal numbers*).

The basic idea is the following. Let $T: [0,1) \longrightarrow [0,1)$, be the transformation defined by $Tx = 10x \pmod 1$ i.e. multiplication by 10, modulo unity. The Lebesgue measure m on this interval is T-invariant and ergodic. Let $B = [0,1/10)$ then it is easy to see that $x_i = 0$ if and only if $T^i x \in B$. Therefore the result follows by applying the above corollary.

(b) Assume $0 < x < 1$ has a continued fraction expansion of the following form

$$x = \cfrac{1}{a_1 + \cfrac{1}{a_2 + \cfrac{1}{a_3 + \cdots}}}, \quad \text{where } a_1, a_2, a_3, \cdots \geq 1.$$

The nth convergent is defined to be

$$p_n / q_n = \cfrac{1}{a_1 + \cfrac{1}{a_2 + \cdots \atop \cdots + \frac{1}{a_n}}}$$

and this sequence of rational numbers satisfies $p_n / q_n \longrightarrow x$, as $n \longrightarrow \infty$. For almost all x (with respect to Lesbegue measure) we have

(i) $\displaystyle \lim_{n \to +\infty} (a_1 \cdots a_n)^{1/n} = \prod_{k=1}^{+\infty} \left(1 + \frac{1}{k^2 + 2k}\right)^{\log k / \log 2}$, and

(ii) $\displaystyle \lim_{n \to +\infty} \frac{\log q_n}{n} = \frac{\pi}{12 \log 2}$.

The ***basic idea*** is the following. Let $T: (0,1) \longrightarrow (0,1)$, be the transformation defined by $Tx = 1/x$ (modulo unity), with the T-invariant measure m being the Gauss measure we defined before. The T-invariant measure m is ergodic. Observe that $a_i = k$ if and only if $T_i x \in [1/(k+1), 1/k)$. Let $f:(0,1) \longrightarrow \mathbb{R}$ be the function defined by setting $f(x) = \log k$ whenever $x \in [1/(k+1), 1/k)$. By the Birkhoff ergodic theorem

$$\frac{1}{n} \sum_{i=1}^{n} \log(a_i) = \frac{1}{n} \sum_{i=0}^{n-1} f(T^i x) \longrightarrow \frac{1}{\log 2} \int_0^1 \frac{f(x)}{1+x} dx = \sum_{k=1}^{+\infty} \frac{\log k}{\log 2} \log\left(1 + \frac{1}{k^2 + 2k}\right)$$

as $n \longrightarrow +\infty$. Taking exponentials of each side of this expression gives a proof of claim (i). The proof of (ii) is similar.

1.6 Proof of the ergodic theorem (Theorem 1.2).

The proof we give is based on a short proof of Katznelson and Weiss [Ktn-Wei].

Preliminary observations. We begin with three observations which make our task easier.

(i) Since any $f \in L^1(X,\mathcal{B},m)$ can be written as $f = f_1 - f_2$, where $f_1, f_2 \geq 0$ are positive functions, we can assume (without loss of generality) that $f \geq 0$.

(ii) We can introduce new functions

$$f^+(x) = \overline{\lim_{n \to +\infty}} \frac{1}{n} \sum_{j=0}^{n-1} f(T^j x) \text{ and } f^-(x) = \underline{\lim_{n \to \infty}} \frac{1}{n} \sum_{j=0}^{n-1} f(T^j x), \text{ for } x \in X,$$

and we have the trivial inequality $f^- \leq f^+$.

Since we can write

$$\frac{1}{n} \sum_{j=0}^{n-1} f(T^j(Tx)) = \left(\frac{n+1}{n}\right) \left(\frac{1}{n+1} \sum_{j=0}^{n} f(T^j x) - \frac{f(x)}{n+1} \right)$$

by taking limits as $n \longrightarrow +\infty$ we have the identities $f^+(Tx) = f^+(x)$, $f^-(Tx) = f^-(x)$ for almost all $x \in X$ (i.e. the functions f^+ and f^- are both T-invariant a.e.). Since m is ergodic and we can apply the lemma on characterizing ergodic measures to deduce that f^-, f^+ are both constant functions a.e.(m).

(iii) To prove the theorem it suffices to show that

$$f^+ \leq \int f \, dm \leq f^- \quad \text{a.e.}(m) \qquad (*)$$

since this is the reverse inequality to $f^- \leq f^+$ (from part (i)) and then $f^+ \equiv f^- \equiv \int f \, dm$.

We now begin the proof in earnest.

Step 1. Choose $\epsilon>0$, then we can define a measurable function n: $X \longrightarrow \mathbb{Z}^+$ by

$$n(x) = \inf\left\{ n\geq 1 \mid f^+ \leq \tfrac{1}{n}\sum_{j=0}^{n-1} f(T^j x) + \epsilon \right\}, \text{ for } x\in X$$

(which is clearly finite on a set of measure one). In particular, we have by definition that

$$n(x)\, f^+ \leq \sum_{j=0}^{n(x)-1} f(T^j x) + n(x)\, \epsilon, \text{ for a.a.}(m)\ x\in X \tag{1.1}$$

Step 2. We want to replace the function $n(x)$ by a uniformly bounded function. For $N>0$ we denote $A = \{x\in X \mid n(x) > N\}$ where we choose N sufficiently large that

$$m(A) < \frac{\epsilon}{f^+} \tag{1.2}$$

We now define new functions

$$\hat{f}(x) = \begin{cases} f(x) & \text{if } x\notin A \\ \max\{f(x), f^+\} & \text{if } x\in A \end{cases} \quad \text{and} \quad \hat{n}(x) = \begin{cases} n(x) & \text{if } x\notin A \\ 1 & \text{if } x\in A \end{cases}$$

The function $\hat{n}$ is uniformly bounded, since by definition $\hat{n}(x)\leq N$ for all $x\in X$. Thus, we have the inequality

$$\hat{n}(x)\, f^+ \leq \sum_{j=0}^{\hat{n}(x)-1} \hat{f}(T^j x) + \hat{n}(x)\, \epsilon \tag{1.3}$$

for a.a.(m) $x\in X$. (Inequality (1.3) follows by (1.1) if $x\notin A$, and is trivial if $x\in A$.)

Observe from (1.2) and the definition of $\hat{f}|_A$ that

$$\int \hat{f}\, dm \leq \int f\, dm + m(A).\, f^+ \leq \int f\, dm + \epsilon \tag{1.4}$$

Step 3. Choose $L>0$ sufficiently large that

$$\frac{N f^+}{L} < \epsilon \tag{1.5}$$

For almost all $x\in X$ we can now define inductively a sequence $n_k(x) \in \mathbb{Z}^+$, $k\geq 0$, as follows.

$$\begin{aligned} n_0(x) &= 0, \\ n_1(x) &= n_0(x) + \hat{n}(T^{n_0(x)}x) \\ n_2(x) &= n_1(x) + \hat{n}(T^{n_1(x)}x) \\ &\vdots \\ n_k(x) &= n_{k-1}(x) + \hat{n}(T^{n_{k-1}(x)}x), \text{ etc.} \end{aligned} \tag{1.6}$$

Given these sequences we can now define a measurable function $k: X \longrightarrow \mathbb{Z}^+$ by

$$k(x) = \sup\{ k \in \mathbb{Z}^+ \mid n_k(x) \leq L-1\}, \text{ for almost all } x \in X.$$

Observe that since $\hat{n}(x) \leq N$ we have $n_k - n_{k-1} \leq N$, $\forall k \geq 2$ and, in particular,

$$L - n_{k(x)}(x) \leq N \tag{1.7}$$

Therefore, for a.a.(m) $x \in X$ we have the equality

$$L f^+(x) \leq \sum_{k=1}^{k(x)} f^+(x) \left(n_k(x) - n_{k-1}(x) \right) + f^+(x) \left(L - n_{k(x)}(x) \right)$$

$$\leq \sum_{k=1}^{k(x)} \left(\sum_{j=n_{k-1}(x)}^{n_k(x)-1} \hat{f}(T^j x) + \hat{n}(T^{n_{k-1}(x)})\epsilon \right) + f^+ N$$

(The first expression is estimated by (1.3), with $T^{n_{k-1}}(x)$ replacing x in each term, and the last term is estimated by (1.7))

$$\leq \sum_{j=0}^{L-1} \hat{f}(T^j x) + L\epsilon + f^+ N \tag{1.8}$$

(We have used the positivity $\hat{f} \geq f \geq 0$ in completing the summation appearing in the first term of (1.8). To bound the second term we have used that $\sum_{k=1}^{k(x)} \hat{n}(T^{n_{k-1}(x)} x) = n_{k(x)}(x)$ is bounded by L, by definition of $k(x)$.)

Step 4. Since $m \in \mathcal{M}_{inv}$ we have the identity $\int \hat{f}(T^j x)\, dm(x) = \int \hat{f}(x)\, dm(x)$, by Lemma 1.1. By integrating both sides of the inequality

(1.8) we get that

$$L f^+(x) \le L \int \hat{f}\, dm + L\epsilon + f^+(x)\ (N-1), \quad a.a.(m)\ x \in X$$

However, we can now use (4) and (5) to get the upper bound

$$\begin{aligned} f^+ &\le \int \hat{f}\, dm + \epsilon + f^+ \frac{(N-1)}{L} \\ &\le \left(\int f dm + \epsilon \right) + \epsilon + \epsilon \quad \text{a.e.}(m) \end{aligned}$$

Since $\epsilon>0$ can be arbitrarily small this proves the first inequality in (*), i.e. $f^+ \le \int f dm$ a.e. The second inequality, $\int f dm \le f^-$ a.e., is similarly proved. □

1.7 Proof of the ergodic decomposition lemma (Lemma 1.4).

We shall give a brief outline proof of the Lemma 1.4 on the ergodic decomposition of invariant measures based on an approach in [Mañé$_2$], pp.162-172, but written in terms of conditional expectations.

Let $\mathfrak{I} = \{A \in X \mid T^{-1}A = A\}$ be the invariant σ-algebra and let m be a T-invariant measure, then we recall the following useful fact. For each $f \in L^2(X,\mathfrak{B},m)$ there exists a projection $\mathrm{E}_m(f|\mathfrak{I}) \in L^2(X,\mathfrak{I},m)$ such that

(i) $\int_A f dm = \int_A \mathrm{E}_m\, (f|\mathfrak{I})\, dm, \ \forall\ A \in \mathfrak{I}$

(ii) $\mathrm{E}_m\ (1|\mathfrak{I}) = 1$

(iii) $\mathrm{E}_m\ (\chi_A|\mathfrak{I}) = \chi_A,\ \forall\ A \in \mathfrak{I}.$

$E_m(f|\mathfrak{I})$ is called the *conditional expectation*, and the above facts can be found in [Parry], pp.20-21, or [Parthasarathy], p.225.

For almost all $x \in X$ (with respect to m) we can define a measure m_x on X by $\int f dm_x = \mathrm{E}_m(f|\mathfrak{I})(x)$, $f \in L^1(X,\mathfrak{B})$. By (ii) above each such measure m_x is a probability measure. To establish ergodicity of the measures m_x we note that if $A \in \mathfrak{L}$ then:

$$\begin{aligned} m_x(A) &= \int \chi_A \, dm_x = E(\chi_A|\mathcal{I})(x) \\ &= \chi_A(x) \qquad \text{(by (iii))} \\ &= \begin{cases} 0 \text{ if } x \in A \\ 1 \text{ if } x \in X - A \end{cases} \end{aligned}$$

By (i) we can write $\int f \, dm = \int_{x \in X} (\int f \, dm_x) \, dm(x)$.

To complete the proof we want to interpret m as a measure on $\mathcal{M}_{erg}$. The measures m_x are indexed by the space X, but X can be embedded into the space $\mathcal{M}_{erg}$ by $X \longrightarrow \mathcal{M}_{erg}$, $x \mapsto m_x$. Therefore, we can interpret m as a measure on $\mathcal{M}_{erg}$ such that

$$m(m') = \begin{cases} 1 & \text{if } m' = m_x (x \in X) \\ 0 & \text{otherwise} \end{cases} \qquad \square$$

Notes

The material in this section is very standard in Ergodic theory. The basic properties of ergodic and invariant measures are well described in many textbooks: for example, **[Walters]**, Chapter 1.

The version of the proof of ergodic decomposition we describe is adapted from Mañé's book **[Mañé$_2$]** (see pp.170-171).

The proof of the Birkhoff ergodic theorem (Theorem 1.1) is taken from an article of Katznelson and Weiss **[Ktn-Wei]**. This has the advantage of appearing more direct than other proofs known to the author. Previously, the most popular proof used the maximal ergodic lemma (see **[Parry]**, pp.24-25). The application of the Birkhoff ergodic theorem to continued fractions is fairly standard and is described in **[Billingsley]**, pp.40-50, or **[Co-Fo-Si]**, pp.174-177, for example.

Chapter 2

Ergodic theorems for manifolds and Liapunov exponents

In the previous chapter we considered the Birkhoff ergodic theorem which: (a) dealt with general measure spaces; and (b) involved limits for averages of general integrable functions.

Henceforth, we shall concentrate on the special case of C^1 diffeomorphisms of compact manifolds. Since we have a differentiable transformation, it is appropriate to prove ergodic theorems in which the integrable function is an expression involving derivatives $D_x f$. More specifically, if we differentiate n-fold compositions of the diffeomorphism then by the chain rule we get the identity $D_x(f^n) = (D_{f^{n-1}x}f)\cdots(D_{fx}f).(D_x f)$, where we write $f^n = f\circ\cdots\circ f$ for the n-fold composition.

It therefore seems natural to expect there should be ergodic theorems to describe this product. In this chapter we shall show that this is indeed the case.

2.1 The subadditive ergodic theorem.

We begin with a preliminary type of 'multiplicative ergodic theorem' in which we get information on the growth of the norms $\|D_x f^n\|$ as $n \longrightarrow +\infty$.

Proposition 2.1 Let $f: M \longrightarrow M$ be a C^1 diffeomorphism of a compact Riemannian manifold and let $m \in \mathcal{M}_{erg}$ be any ergodic measure. Then there exists $\lambda \in \mathbb{R}$ such that $\lim_{n\to+\infty} \frac{1}{n}\log\|D_x f^n\| = \lambda$, for almost all $x \in M$.

Remark. Since M is a C^1 Riemannian manifold there is a natural inner product on each fiber $T_x M$, $x \in M$, of the tangent space TM. The norm $\|D_x f^n\|$ of the linear map $D_x f^n$: $T_x M \longrightarrow T_{f^n x} M$ is relative to the norms on $T_x M$, $T_{f^n x} M$ coming from the inner product on these two fibers. If we choose an equivalent Riemannian metric for M we would get exactly the same limit λ, since the expression $\log\|D_x f^n\|$ will only change by a bounded quantity and thus the difference in $\frac{1}{n}\log\|D_x f^n\|$ will disappear as $n \longrightarrow +\infty$.

As we shall show in Section 2.2, Proposition 2.1 is an immediate corollary of the following type of general ergodic theorem.

Proposition 2.2 (Subadditive ergodic theorem) Let $(X,\mathcal{B})$ be a measurable space with a measurable transformation $T: (X,\mathcal{B}) \longrightarrow (X,\mathcal{B})$ and let $m \in \mathcal{M}_{erg}$ be any ergodic measure. Let $(F_n)_{n=1}^{+\infty} \in L^1(X,\mathcal{B},m)$ be a sequence of functions such that,

$$\forall n,k \geq 1,\ F_{n+k}(x) \leq F_n(x) + F_k(T^n x)\ ,\ \text{for a.a.}(m)\ x \in X$$

(i.e. the *subadditivity condition*). Then,

(i) $\exists \lambda \in \mathbb{R} \cup \{-\infty\}$ such that $\lim_{n \to +\infty} \frac{1}{n} F_n(x) = \lambda$, for almost all $x \in X$,

(ii) the value λ is given by $\lambda = \inf \{ \frac{1}{n} \int F_n \, dm \mid n \geq 1 \}$.

The proof of Proposition 2.2 is somewhat technical and so we prefer to postpone it until later in this chapter.

2.2 The subadditive ergodic theorem and diffeomorphisms.

We shall now show how to derive Proposition 2.1 from Proposition 2.2.

Assuming the hypothesis of Proposition 2.1 we want to apply Proposition 2.2 where we take X to be the manifold and $\mathcal{B}$ to be the Borel σ-algebra. In Proposition 2.2 we set $F_n(x) = \log\|D_x f^n\|$, $n \geq 1$. (Notice that $D_x f^n$ is bounded, and thus F_n is integrable, since M is compact and each map $x \mapsto D_x f^n$ is continuous.) By the chain rule we have the inequality

$$\left\| D_x f^{(n+k)} \right\| = \left\| (D_{f^n x} f^k) . (D_x f^n) \right\| \leq \left\| D_{f^n x} f^k \right\| . \left\| D_x f^n \right\| .$$

By taking logarithms of each side we see that the subadditivity condition holds for F_n. Finally, applying Proposition 2.2 gives that $\lim_{n \to +\infty} \frac{1}{n} F_n(x) = \lim_{n \to +\infty} \frac{1}{n} \log \|D_x f^n\| = \lambda$, for almost all $x \in M$, which completes the derivation of Proposition 2.1. □

Remark. At the level of generality of Proposition 2.2 (i) we have to allow the possibility that $\lim_{n \to +\infty} \frac{1}{n} F_n(x) = -\infty$. However, in our application to the proof of Proposition 2.1 we can use the bounds

$$\frac{1}{\|Df^{-1}\|^n} \leq \frac{1}{\left\|D_{f^n x}(f^{-1})^n\right\|} \leq \|D_x f^n\|,$$

where $\|Df^{-1}\| = \sup\left\{\log\|D_x f^{-1}\| \mid x \in M\right\}$, etc., to estimate that $-\log\|Df^{-1}\| \leq \frac{1}{n}F_n(x) = \frac{1}{n}\log\|D_x f^n\|$, $\forall n \geq 1$. In particular, this sequence of functions is uniformly bounded below, and the limit λ must be finite.

2.3 Oseledec-type theorems.

If M is a d-dimensional manifold then the linear tangent map $D_x f^n$: $T_x M \longrightarrow T_{f^n x} M$ can be represented by an $d \times d$ matrix (with respect to any appropriate basis for the two d-dimensional vector spaces). Proposition 2.1 gives information about the growth rate of the *norm* of this matrix as $n \longrightarrow +\infty$. There are sharper results, originally due to Oseledec, about the growth rate of $\|D_x f^n(v)\|$ for *individual* vectors $v \in T_x M$ as $n \longrightarrow +\infty$.

The basic form of these results is well illustrated by the (notationally) simpler version for surfaces (i.e. when $d=2$). Therefore, we begin by stating a version of the result of Oseledec in that context.

Theorem 2.1 (Oseledec, surfaces). Let f: $M \longrightarrow M$ be a C^1 diffeomorphism of a compact surface and let $m \in \mathcal{M}_{erg}$ be any ergodic measure. Then there are two possibilities: *either*

(i) there exists $\lambda \in \mathbb{R}$ such that $\forall\ v \in T_x M$,

$$\lim_{n \to +\infty} \frac{1}{n} \log \|D_x f^n(v)\| = \lambda$$

(for almost all $x \in M$) , *or*

(ii) there exist $\lambda_1 > \lambda_2$ and a splitting $T_x M = E_x^1 \oplus E_x^2$ (with the maps $x \mapsto E_x^1$, E_x^2 being measurable) such that

$$\begin{cases} \text{for } v_1 \in E_x^1, \lim_{n \to +\infty} \frac{1}{n}\log\|Df^n(v_1)\| = \lambda_1 \\ \text{for } v_2 \in E_x^2, \lim_{n \to +\infty} \frac{1}{n}\log\|Df^n(v_2)\| = \lambda_2 \end{cases}$$

for a.a.(m) $x \in X$

We shall present a proof of the above result in the next section.

Definition. For a given C^1 diffeomorphism $f: M \longrightarrow M$ the numbers λ or λ_1, λ_2 occurring in the statement of the Oseledec theorem are called the *Liapunov exponents* of the measure $m \in \mathcal{M}_{erg}$.

Remarks. (i) The largest Liapunov exponent λ_1 (or λ) will correspond to the rate of growth of $\|D_x f^n\|$, which was denoted by λ in the statement of Proposition 2.1.

(ii) If f is replaced by f^{-1} then any ergodic measure $m \in \mathcal{M}_{erg}$ for f is again an ergodic measure for f^{-1}. However, in the Oseledec theorems the Liapunov exponents $\lambda_1 \geq \lambda_2$ for f are replaced by Liapunov exponents $-\lambda_2 \geq -\lambda_1$ (This will soon be seen in Step 1, Section 2.5). Thus in view of remark (i) we see that

$$\lambda_1 = \lim_{n \to +\infty} \tfrac{1}{n} \log \|D_x f^n\|, \text{ and}$$

$$\lambda_2 = -\lim_{n \to +\infty} \tfrac{1}{n} \log \|D_x f^{-n}\|$$

(for a.a.(m) $x \in M$).

(iii) Consider the special case of a diffeomorphism $f: M \longrightarrow M$ for which the (normalized) volume is an f-invariant probability measure m and, furthermore, assume m is ergodic. Since f is volume preserving we know that the Jacobian $\mathrm{Jac}(f)$ is identically 1 and therefore $e^{\lambda_1} e^{\lambda_2} = \mathrm{Det}(Df^n) = \mathrm{Jac}(f^n) \equiv 1$. We therefore conclude that for volume-preserving transformations, and the normalised volume m, we have $\lambda_1 = -\lambda_2$.

In these notes we shall only need the above version of Oseledec's theorem for surfaces. However, for completeness we mention that there exists a more general version of this theorem for dimensions $d \geq 2$ and it takes the form stated below.

Theorem 2.2 (Oseledec, arbitrary dimension). Let $f: M \longrightarrow M$ be a C^1 diffeomorphism of a compact manifold of dimension n and let $m \in \mathcal{M}_{erg}$. Then there exist:

(a) real numbers $\lambda_1 > \cdots > \lambda_k$ $(k \leq n)$;

(b) positive integers $n_1, \cdots, n_k \in \mathbb{Z}^+$ such that $n_1 + \cdots + n_k = n$;

(c) a measurable splitting $T_x M = E_x^1 \oplus \cdots \oplus E_x^k$, with $\dim(E_x^i) = n_i$ and

$D_x f(E_x^i) = E_{fx}^i$, such that

$$\lim_{n\to+\infty} \tfrac{1}{n}\log\|D_x f^n(v)\| = \lambda_\ell \text{ (for a.a.}(m)\ x\in M),$$

whenever $v\in(E_x^1\oplus\cdots\oplus E_x^\ell)$, but $v\notin(E_x^1\oplus\cdots\oplus E_x^{\ell-1})$.

These numbers $\lambda_1,\cdots,\lambda_k$ are again called the *Liapunov exponents* of the ergodic measure m.

Remarks. (i) It is easy to manufacture examples with a non-zero Liapunov exponent. If ϕ_t: $M\longrightarrow M$ is a C^1 flow then we define the time-T transformation $f: M\longrightarrow M$ by $f(x) = \phi_T(x)$, for some fixed real number T. Any ϕ-invariant ergodic measure m will be an f-invariant ergodic measure with a Liapunov exponent which is zero (corresponding to the flow direction for ϕ).

(ii) If we take determinants we can apply the Birkhoff ergodic theorem (Theorem 1.2) to see that for any $m\in\mathcal{M}_{\text{erg}}$

$$\tfrac{1}{n}\log|D_x(f^n)| = \lim_{n\to+\infty}\tfrac{1}{n}\sum_{j=0}^{n-1}(D_{f^j x} f) = \int\log|D_x f|, \text{ a.a.}(m)\ x\in M.$$

It is not difficult to deduce from the Oseledec theorem that this limit equals the sum of the positive Liapunov exponents $\left(\sum_{\lambda_i>0}\lambda_i\right)$.

2.4 Some examples.

The Oseledec theorem is better understood after we think about a few examples.

The trivial examples. (i) ('Either' case) Let $M = \mathbb{R}^2/\mathbb{Z}^2$ be the standard flat torus and consider the transformation $f: M\longrightarrow M$ defined by $f(x_1,x_2)+\mathbb{Z}^2 = (x_1+\alpha_1,x_2+\alpha_2)+\mathbb{Z}^2$. We can consider the ergodic measure m which is the usual Lebesgue-Haar measure on M. Since the derivative at any point $x\in M$ takes the form $D_x f \equiv \left[\begin{smallmatrix}1&0\\0&1\end{smallmatrix}\right]$ we see that $D_x f^n(v) = v$, $\forall n\geq 1$, $\forall v\in T_x M$. Therefore, we trivially have

$$\lim_{n\to+\infty} \frac{1}{n}\log\|D_x f^n(v)\| = \lim_{n\to+\infty}\frac{1}{n}\log\|v\|\equiv 0, \quad \forall v\in T_x M$$

and so $\lambda_1=\lambda_2=0$.

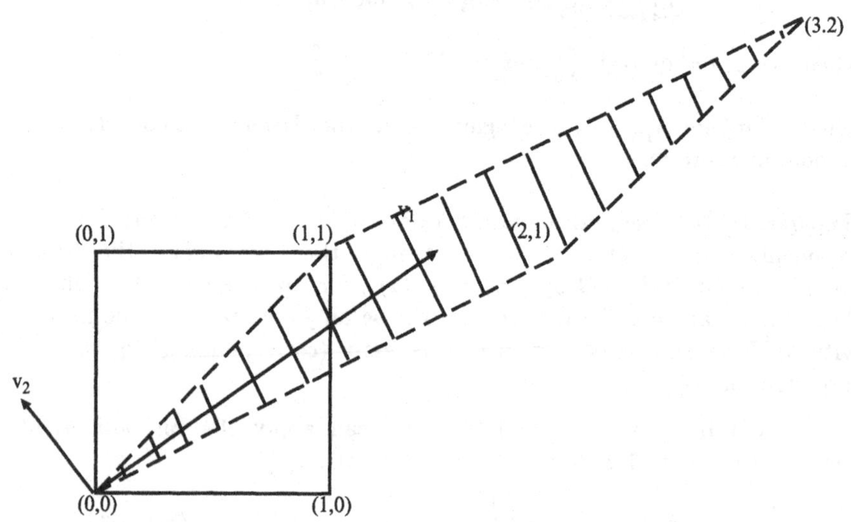

Figure 4: Distinct Liapunov exponents on the torus

(ii) ('Or' case) Again we let $M= \mathbb{R}^2/\mathbb{Z}^2$ be the standard flat torus, but this time we define the diffeomorphism $f: M \longrightarrow M$ by $f(x_1,x_2)+\mathbb{Z}^2=(2x_1+x_2,x_1+x_2)+\mathbb{Z}^2$. Again the Lebesgue-Haar measure m on M is ergodic. For any point $x \in M$ the derivative takes the form $D_x f \equiv \begin{bmatrix} 2 & 1 \\ 1 & 1 \end{bmatrix}$. This matrix has the two eigenvalues $(3 \pm \sqrt{5})/2$ and we denote the associated eigenvectors by v_1, v_2. Thus we easily see that the Liapunov exponents are

$$\lambda_1 = \log\left(\frac{3+\sqrt{5}}{2}\right) \text{ and } \lambda_2 = \log\left(\frac{3-\sqrt{5}}{2}\right)$$

where the splitting $E_x^1 \oplus E_x^2$ is a 'translate' of the splitting at the origin $x=(0,0)$ given by the eigenspaces spanned by v_1 and v_2 (Figure 4).

(iii) ('Or' case) Let R be a rectangle in $\mathbb{R}^2$ and define $f: \mathbb{R}^2 \longrightarrow \mathbb{R}^2$ so that it maps the rectangle R as shown in Figure 5.

We assume that the map f extends to a diffeomorphism $f: S^2 \longrightarrow S^2$ of the standard 2-sphere (by adding a fixed repelling point at infinity). The map f is assume to contract the rectangle (on $R \cap f^{-1}R$) in the horizontal direction by $\beta = 1/\alpha < 1/2$, say, and expand the rectangle (on $R \cap fR$) in the vertical direction by $\alpha > 2$. This map is usually called the *Smale horse-shoe.* Let $\Omega = \bigcup_{n=-\infty}^{+\infty} f^{-n} R$ be the set of points in R which remain in R under all forward and backward iterates. The set Ω is homeomorphic to a Cantor set by an obvious bijection with $\{0,1\}^{\mathbb{Z}}$.

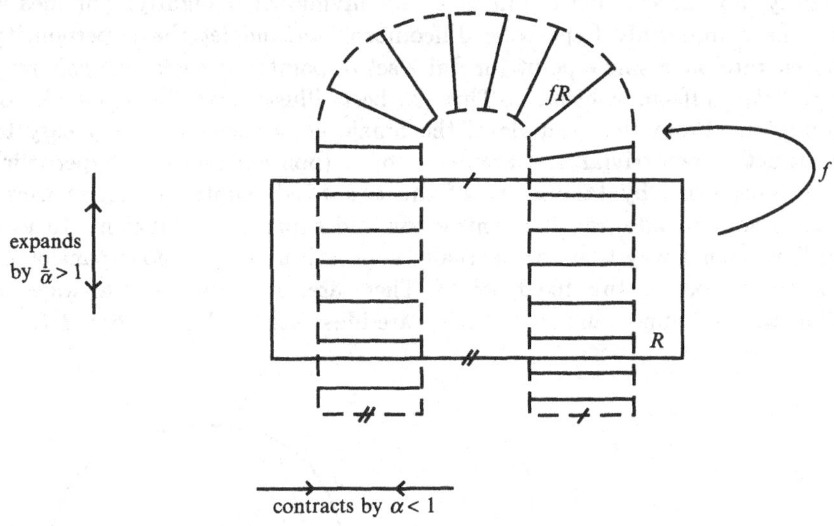

Figure 5: Smale horse-shoe

Let $m = (\frac{1}{2}, \frac{1}{2})^{\mathbb{Z}}$ be the natural Bernoulli measure supported on Ω (i.e. the measure on Ω such that each component of $\bigcap_{n=-p}^{+q} f^{-n} R$ has measure $(\frac{1}{2})^{p+q+1}$ then m is ergodic.

For any point $x \in \Omega$ the derivative of f takes the form $D_x f \equiv \begin{bmatrix} \alpha & 0 \\ 0 & 1/\alpha \end{bmatrix}$. Therefore, it is easy to check that $\exp(\lambda_1) = 1/\alpha$, $\exp(\lambda_2) = \alpha$ and the splitting $T_x M = E_x^1 \oplus E_x^2$ corresponds to the horizontal (contracting) directions and the vertical (expanding) directions.

Remark. The above examples have the advantage that they are very simple and the Liapunov exponents can be easily computed. However, this is *not* generally true. These particular examples have the exceptional properties that the limiting behavior for *each* point $x \in \mathrm{supp}(m)$ is the same, and thus it is not really necessary to use the Oseledec theorem to study them. This property is characterisitic of *uniformly hyperbolic* diffeomorphisms (see Interlude, Part(b)).

More interesting examples. The above examples are perhaps a little *too* simple. We now want to describe a general method of constructing slightly more interesting examples by modifying them slightly. The idea is to take a uniformly hyperbolic diffeomorphism and let the hyperbolicity 'degenerate' at a single point (or finite set of points) to get a *non-uniformly hyperbolic* diffeomorphism. This is best illustrated by a couple of examples. From the example of the Smale horse-shoe, it is very easy to construct non-trivial examples of (non-uniformly hyperbolic) diffeomorphisms by taking one of the two fixed points in the invariant Cantor set and allowing the contraction and expansion coefficients to *vary* on R in such a way that they actually approach unity (i.e. no expansion or no contraction) at this fixed point. There are, of course, several ways to allow this to happen and two of these are illustrated in Figures 6 and 7.

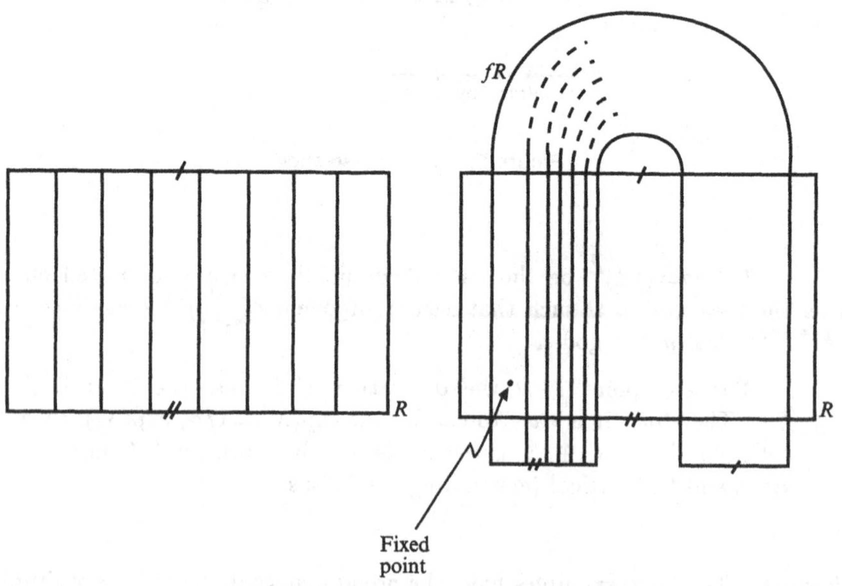

Figure 6: Contraction tends to unity at a fixed point

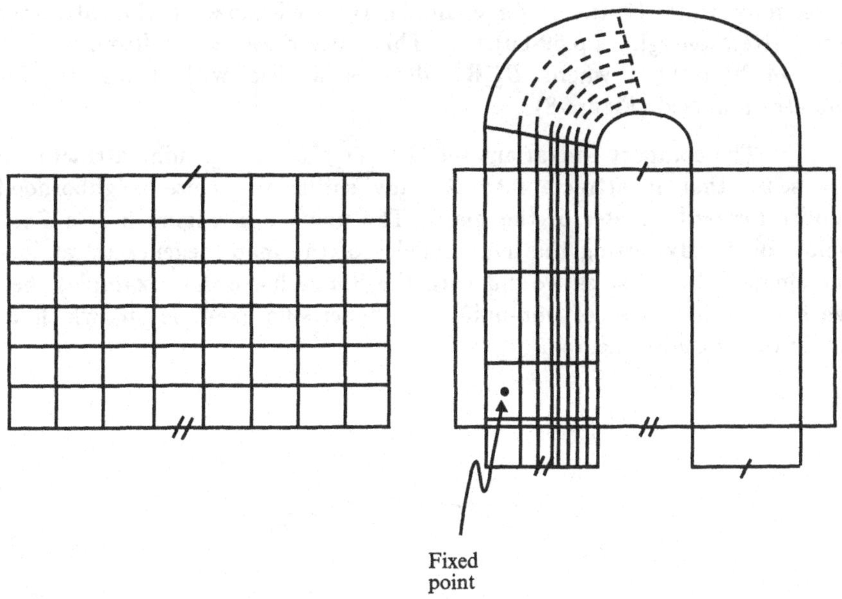

Figure 7: Contraction and expansion tends to unity at a fixed point

(a) (A modified horse-shoe) Figure 6 shows an example where the contraction rate $\alpha(x)$ approaches unity at a fixed point (but the expansion rate $\beta(x) \geq \beta > 1$ remains bounded away from unity).

(b) (Another modified horse-shoe) Figure 7 shows where both the contraction rate $\alpha(x) < 1$ and the expansion rate $\beta(x) > 1$ approach unity at a fixed point.

In each of these cases we can consider the invariant Cantor set $\Omega = \bigcap_{n=-\infty}^{+\infty} f^n R$, and the invariant (ergodic) measure $m = (\frac{1}{2}, \frac{1}{2})^{\mathbb{Z}}$ on Ω. By the Oseledec theorem we have Liapunov exponents which describe the asymptotic behavior at *almost all points* (with respect to the measure m) and it is not difficult to see that these exponents must be non-zero.

However, we see that in these examples the set of zero measure which we are at liberty to ignore is non-empty since it includes the 'degenerate' fixed point where we have arranged that there is *no hyperbolicity whatsoever.* Thus these examples begin to hint at the strength of the Oseledec theorem as a generalization beyond the uniformly hyperbolic cases.

It is easy to see that this procedure can be applied to other surface diffeomorphisms. Consider, for example, the well-known Plykin attractor (see [**Newhouse**$_3$], pp.59-60). This describes a diffeomorphism $f: D \longrightarrow f(D) \subseteq \text{int}(D)$, where $D \subseteq \mathbb{R}^2$ denotes a disc with three disjoint subdiscs removed (Figure 8).

The compact f-invariant set $\Omega = \bigcap_{n=0}^{+\infty} f^n D$, is a genuine attractor in the sense that it attracts into Ω any sufficiently close neighborhood (under forward iterates of the map). The point x in Figure 8 is a fixed point for f. By letting the hyperbolicity of the map 'degenerate' at just the single point (just as we did with the Smale horse-shoe example) then we get another class of non-uniformly hyperbolic examples to which to apply the Oseledec theorem.

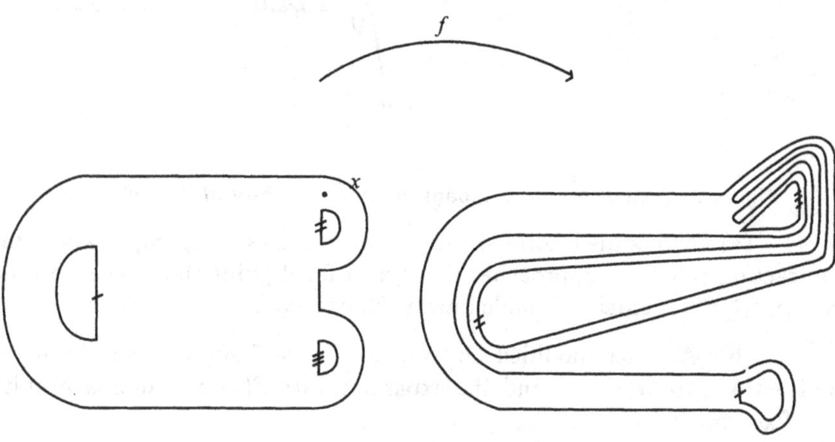

Figure 8: The Plykin attractor

Finally, it is worth noting that, given any surface M and point $x \in M$, we can construct in a trivial way a diffeomorphism $f: M \longrightarrow M$ with $f(x) = x$ and with the derivative at this fixed point being given by

$$D_x f = \begin{bmatrix} e^{\lambda_1} & 0 \\ 0 & e^{\lambda_2} \end{bmatrix} \quad \text{for any choice } \lambda_1 > \lambda_2.$$

If we choose our ergodic measure $m = \delta_x$ to be the Dirac measure at x then λ_1, λ_2 are the Liapunov exponents for m. Clearly, we can repeat this construction for any finite number of distinct points $x_1, \cdots, x_n$ and pairs $\lambda_1^{(1)}$, $\lambda_2^{(1)}, \cdots,$ $\lambda_1^{(n)}$, $\lambda_2^{(n)}$ (Figure 9). In particular, we can then conclude that if $m_i = \delta_{x_i}$ is the Dirac measure supported on the fixed point x_i then the associated Liapunov exponents are $\lambda_1^{(i)}$, $\lambda_2^{(i)} (1 \leq i \leq n)$

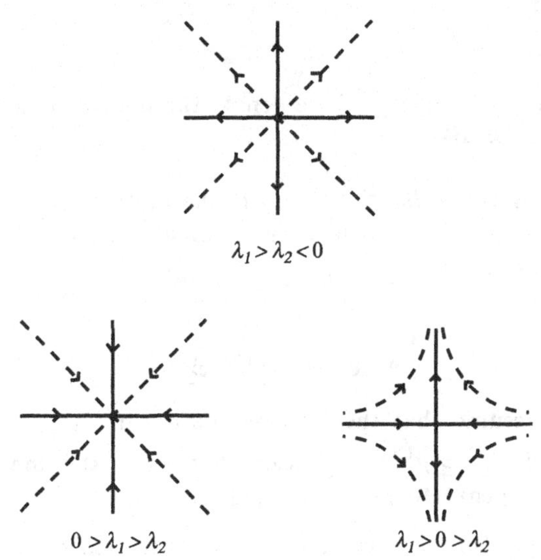

Figure 9: Liapunov exponents at fixed points.

2.5 Proof of the Oseledec theorem.

In this section we shall present a proof of Oseledec's theorem for surfaces. The proof of the theorem for arbitrary dimensions is similar in spirit, but technically more complicated. We shall follow the method of proof used in [Ruelle$_1$], pp.31-32.

We shall present the proof in a sequence of steps.

Step 1. The basic idea is to construct the Liapunov exponents from the eigenvalues of certain symmetric matrices A_n associated to the matrices B_n representing $D_x f^n$, $n \geq 1$. The advantage of this substitution is that symmetric matrices always have real eigenvalues, whilst this may not necessarily always be the case with B_n. We require the following standard lemma.

Lemma 2.1 (cf. [Rudin], p.278). If B is a square matrix then there exists a (positive) symmetric matrix $A=A^*$ such that:

(i) $A^2 = B^*B$; and

(ii) $\|Av\|=\|Bv\|$.

Notation. In view of property (i) we denote the matrix A associated to the matrix B by $A=(B^*B)^{1/2}$.

Given the 2×2 matrices $B_n \equiv D_x f^n$, $n \geq 1$, the matrix $A_n = (B_n^* B_n)^{1/2}$ is a 2×2 symmetric matrix with real eigenvalues $\mu_1^{(n)}$ and $\mu_2^{(n)}$ satisfying the eigenvalue equations

$$\begin{cases} A_n\, e_1^{(n)} = \mu_1^{(n)}\, e_1^{(n)} \\ A_n\, e_2^{(n)} = \mu_2^{(n)}\, e_2^{(n)} \end{cases}$$

where we may assume that the eigenvectors are normalized i.e. $\left\|e_1^{(n)}\right\| = \left\|e_2^{(n)}\right\| = 1$ and $\mu_1^{(n)} \geq \mu_2^{(n)}$. (Notice that since the matrices A_n are symmetric these eigenvectors are orthogonal.)

By Proposition 2.1 we know that there exists $\lambda \in \mathbf{R}$ such that $\lim_{n\to+\infty} \frac{1}{n}\log\|B_n\| = \lambda$ a.e., and since by part (ii) of Lemma 2.1 we know that $\|A_n\|=\|B_n\|$ we therefore have

$$\lambda = \lim_{n\to+\infty} \tfrac{1}{n}\log\|B_n\| = \lim_{n\to+\infty} \tfrac{1}{n}\log\|A_n\| = \lim_{n\to+\infty} \tfrac{1}{n}\log\mu_1^{(n)}. \tag{2.1}$$

Similarly, if we replace the diffeomorphism $f: M \longrightarrow M$ by its inverse $f^{-1}: M \longrightarrow M$ and carry out the above argument we get that there exists $\nu \in \mathbf{R}$ such that

$$\nu = \lim_{n\to+\infty} \tfrac{1}{n}\log\|D_x f^{-n}\|, \quad \text{a.a.}(m)\ x \in X \tag{2.2}$$

i.e. from (2.1) and (2.2) we deduce that there exists real numbers $\lambda_1 = \lambda$ and $\lambda_2 = \nu$ such that (for a.a.(m) $x \in M$),

$$\begin{cases} \lambda_1 = \lim_{n\to+\infty} \frac{1}{n} \log\|D_x f^n\| \\ \lambda_2 = \lim_{n\to+\infty} \frac{1}{n} \log\|D_x f^{-n}\| \end{cases} \tag{2.3}$$

These are the Liapunov exponents λ_1, λ_2 in the statement of Oseledec's theorem.

Step 2. We next want to construct the splitting $T_x M = E_x^1 \oplus E_x^2$ of the tangent space for the case $\lambda_1 \neq \lambda_2$. Without loss of generality, we assume that $\lambda_1 > \lambda_2$.

Let $E_1^{(n)} = \text{span}\{e_1^{(n)}\}$ and $E_2^{(n)} = \text{span}\{e_2^{(n)}\}$ then we want to construct E_x^1 as a limit of the subspaces $\{E_1^{(n)}\}_{n=1}^{+\infty}$ in $T_x M$, for almost all points $x \in M$. By completeness, it suffices to show that this sequence $\{E_1^{(n)}\}_{n=1}^{+\infty}$ is Cauchy.

We begin by comparing the spaces $E_1^{(n)}$, $E_1^{(n+1)}$, $n \geq 1$. Observe that by using the orthonormality of the eigenvectors, we can write

$$1 = \left\| e_1^{(n+1)} \right\|^2 = \left| < e_1^{(n)}, e_1^{(n+1)} > \right|^2 + \left| < e_2^{(n)}, e_1^{(n+1)} > \right|^2$$

and thus

$$\begin{aligned} \left\| e_1^{(n)} - e_1^{(n+1)} \right\| &= \left| < e_1^{(n)} - e_1^{(n+1)},\ e_1^{(n)} - e_1^{(n+1)} > \right|^{1/2} \\ &= \left| 2\left(1 - < e_1^{(n)}, e_1^{(n+1)} > \right) \right|^{1/2} \\ &= \left| 2\left(1 \pm \left\{1 - \left| < e_2^{(n)},\ e_1^{(n+1)} > \right|^2 \right\}^{1/2} \right| ^{1/2} \\ &\leq C \left| < e_2^{(n)}, e_1^{(n+1)} > \right|, \end{aligned}$$

for some constant $C > 0$, providing $\left| < e_2^{(n)}, e_1^{(n+1)} > \right|$ is sufficiently small. In fact, we claim that $< e_2^{(n)}, e_1^{(n+1)} > \to 0$ as $n \to +\infty$ (as we shall show below).

Given $\epsilon > 0$, we know by step 1 that for a.a.(m) $x \in X$, and all sufficiently large $n \geq 1$, $|\lambda_1 - \frac{1}{n} \log \mu_1^{(n)}| < \epsilon$. Therefore, we can estimate

$$\begin{aligned} \left| < e_1^{(n+1)},\ e_2^{(n)} > \right| &= \left| < \frac{A_{n+1} e_1^{(n+1)}}{\mu_1^{(n+1)}},\ e_2^{(n)} > \right| \\ &\leq \frac{\left| < e_1^{(n+1)},\ A_{n+1} e_2^{(n)} > \right|}{e^{(n+1)(\lambda_1 - \epsilon)}} \end{aligned} \tag{2.4}$$

But the expression in the numerator of (2.4) can be estimated by:

$$\begin{aligned} \left|< e_1^{(n+1)}, A_{n+1} e_2^{(n)} >\right| &\le \left\| A_{n+1} e_2^{(n)} \right\| = \left\| B_{n+1} e_2^{(n)} \right\| \\ &\le \left\| \mathrm{D}_{f^n x} f \right\| . \left\| B_n e_2^{(n)} \right\| \\ &= \left\| \mathrm{D}_{f^n x} f \right\| . \left\| A_n e_2^{(n)} \right\| \end{aligned} \tag{2.5}$$

(using Lemma 2.1 (ii) and $B_{n+1} = \mathrm{D}_x f^{n+1} = \mathrm{D}_{f^n x} f \circ B_n$, by the chain rule).

Since f is a C^1 diffeomorphism and M is compact we can define $D = \sup\{\|\mathrm{D}_y f\| \mid y \in M\} < +\infty$. Furthermore, we observe that if n is sufficiently large then

$$\left\| A_n e_2^{(n)} \right\| = \mu_2^{(n)} \le \mathrm{e}^{(\lambda_2 + \epsilon)n}.$$

Using these two estimates and (2.5) the inequalities in (2.4) give

$$\left|< e_1^{(n+1)}, e_2^{(n)} >\right| \le \frac{D}{\mathrm{e}^{(\lambda_1 - \epsilon)}} . \ \mathrm{e}^{-(\lambda_1 - \lambda_2 + 2\epsilon)n},$$

for n sufficiently large.

Since we are dealing with the case $\lambda_1 > \lambda_2$, and because we have the freedom to choose $\epsilon > 0$ as small as we like, we can now assume $0 < \epsilon < (\lambda_1 - \lambda_2)/2$. In view of the above estimate we see that there exists $\delta > 0$ such that, for sufficiently large n, $\left|< e_1^{(n+1)}, e_2^{(n)} >\right| \le \mathrm{e}^{-n\delta}$. This completes the proof of our clain and, furthermore, we get the estimate

$$\left\| e_1^{(n)} - e_1^{(n+1)} \right\| \le C\, \mathrm{e}^{-n\delta} \tag{2.6}$$

for n sufficiently large.

To complete our proof of the Cauchy property we can iterate (2.6), as follows. For any $k \ge 1$,

$$\begin{aligned}\left\|e_1^{(n)}-e_1^{(n+k)}\right\| &\le \left\|e_1^{(n)}-e_1^{(n+1)}\right\|+\cdots+\left\|e_1^{(n+k-1)}-e_1^{(n+k)}\right\| \\ &\le C\left(e^{-n\delta}+e^{-(n+1)\delta}+\cdots+e^{-(n+k)\delta}\right) \\ &\le \frac{C\,e^{-n\delta}}{(1-e^{-\delta})}. \end{aligned} \tag{2.7}$$

The conclusion we draw from (2.7) is that $E_1^{(n)}$ is Cauchy, and therefore has a limit E_1, say. (The bundle E_2 can similarly be constructed by repeating this argument for f^{-1}.)

Step 3. We shall now show that $u\in E_1$ satisfies $\lim_{n\to+\infty}\frac{1}{n}\log\|B_n u\| = \lambda_1$ (for a.a.(m) $x\in M$) . For $u\in E_1$, with $\|u\| = 1$, we can estimate

$$\left|\|B_n u\| - \left\|B_n e_1^{(n)}\right\|\right| \le \left\|B_n(u-e_1^{(n)})\right\|.$$

To bound the right-hand side of this expression we observe that:

(a) if we let $k\to+\infty$ in the inequality (2.7) we can estimate

$$\left\|e_1^{(n)}-u\right\| \le \frac{C\,e^{-n\delta}}{(1-e^{-\delta})}$$

for sufficiently large n;

(b) for sufficiently large n we have $\|B_n\| \le e^{(\lambda_1+\delta/2)n}$ (by Proposition 2.1).

To bound $\left\|B_n e_1^{(n)}\right\|$ we observe that

(c) for all $n>0$, $\left\|B_n e_1^{(n)}\right\| = \left\|A_n e_1^{(n)}\right\| = \mu_1^{(n)}$;

(d) for any $\epsilon'<\frac{\delta}{2}$ we can estimate $e^{(\lambda_1-\epsilon')n}\le\mu_1^{(n)}\le e^{(\lambda_1+\epsilon')n}$, for sufficiently large n.

Putting together all these estimates gives

$$e^{(\lambda_1-\epsilon')n}-\frac{e^{(\lambda_1-\frac{\delta}{2})n}}{(1-e^{-\delta})}\le\|B_n u\|\le e^{(\lambda_1+\epsilon')n}+\frac{e^{(\lambda_1-\frac{\delta}{2})n}}{(1-e^{-\delta})} \tag{2.8}$$

Taking the limit $n\longrightarrow+\infty$ in (2.8) gives:

$$\overline{\lim_{n\to+\infty}} \frac{1}{n} \log\|B_n u\| \leq \lambda_1+\epsilon' \quad \text{and} \quad \underline{\lim_{n\to+\infty}} \frac{1}{n} \log\|B_n u\| \geq \lambda_1-\epsilon'$$

Finally, notice that since $\epsilon'>0$ can be chosen arbitrarily small we can deduce that $\lim_{n\to+\infty} \frac{1}{n} \log\|D_x f^n u\| = \lambda_1$, whenever $u\in E_1$ (for a.a.(m) $x \in M$).

By a similar argument we can show that $\lim_{n\to+\infty} \frac{1}{n}\log\|D_x f^{-n} u\|= \lambda_2$, for $u\in E_2$ (for a.a.(m) $x\in M$)

Step 4. All that remains is to deal with the missing case where $\lambda_1=\lambda_2=\lambda$. In fact, this is even easier since for any $u\in T_x M$ we write $u=\alpha_n e_1^{(n)}+\beta_n e_2^{(n)}$ and then estimate $\|B_n u\|$, as above, in terms of $\left\|B_n e_1^{(n)}\right\|$ and $\left\|B_n e_2^{(n)}\right\|$. □

Remark. The splitting $x\mapsto E_x^1\oplus E_x^2$ is measurable because the construction of these spaces involved taking certain limits, under which the set of measurable functions is closed.

2.6. Further refinements of the Oseledec theorem.

We briefly mention two ways in which the Oseledec theorem can be refined.

(a) The Oseledec theorem, as we have chosen to state it, gives limits on a set of full measure Ω associated to a given ergodic measure $m\in \mathcal{M}_{erg}$. It is possible to reformulate the theorem to say that the splitting and associated limits hold on a single *total probability set* Λ which is independent of the choice $m\in\mathcal{M}_{erg}$. (A total probability set $\Lambda \subseteq M$ has the property that $\mu(\Lambda) = 1$ for <u>all</u> invariant measures $\mu \in\mathcal{M}_{inv}$.)

(b) The theorem can also be generalized to deal with any invariant measure $m\in\mathcal{M}_{inv}$, without necessarily assuming ergodicity. The statement of the theorem has to be modified to the effect that for almost all points $x\in\Lambda$ the Liapunov exponents $\lambda_1,\cdots,\lambda_k$ will be replaced by measurable functions $\lambda_1(x),\cdots, \lambda_k(x)$ depending on the choice of x (rather than just being constants depending only on m, as in the ergodic case).

2.7 Proof of subadditive ergodic theorem (Proposition 2.2).

We now give a proof of Proposition 2.2 we postponed earlier in this chapter. This result was originally due to Kingman, although the proof we give is a more recent proof of Katznelson and Weiss [Ktn-Wei],pp.293-296.

We begin with the simple observation that since $\int F_{n+k} dm \leq \int F_n dm + \int F_k dm$, $\forall n,k \geq 1$, then $\frac{1}{n}\int F_n dm$ converges to the constant $\lambda = \inf\{ \frac{1}{n}\int F_n dm \mid n \geq 1\}$. This uses the following standard lemma.

Lemma 2.2 (Subadditivity of sequences). If $\{a_n\}_{n=1}^{+\infty}$ is a sequence of real numbers satisfying the subadditivity condition $a_{n+m} \leq a_n + a_m$, $\forall n,m \geq 1$, then $\frac{a_n}{n} \longrightarrow \inf\left\{\frac{a_n}{n} \mid n \geq 1\right\}$, as $n \longrightarrow +\infty$.

The proof can be found in [Bowen], p.28.

We denote $F^+(x) = \overline{\lim} \frac{1}{n} F_n(x)$ and $F^-(x) = \underline{\lim} \frac{1}{n} F_n(x)$, $x \in X$. We shall divide the proof of Proposition 2.2 into a sequence of steps.

Step 1. By iterating the subadditivity condition n times

$$\frac{1}{n} F_n(x) \leq \frac{1}{n}\sum_{j=0}^{n-1} F_1(T^j x) \tag{2.9}$$

By the Birkhoff ergodic theorem (Theorem 1.2) we have the inequality

$$F^-(x) \leq \int F_1 \, dm \tag{2.10}$$

and by integrating (2.9), and recalling that the measure m is T-invariant, we get

$$\lambda \leq \frac{1}{n}\int F_n dm \leq \int F_1 dm. \tag{2.11}$$

Step 2. We want to 'improve' the inequality (2.10) as follows. First fix $N>1$, and assume $n>N$. For each $1 \leq i \leq N$ we can make unique choices $m=m(i) \geq 0$, $N>k=k(i) \geq 0$, such that $n=i+(mN+k)$. By the subadditivity condition we can bound $F_n(x)$, for almost all $x \in M$, by

$$F_n(x) \leq F_i(x) + \sum_{\ell=0}^{m(i)-1} F_n(T^{\ell N+i} x) + F_k(T^{mN+i} x) \tag{2.12}$$

We can sum both sides of (2.12) over $1 \leq i \leq N$ to get

$$
\begin{aligned}
NF_n(x) &\leq \sum_{i=1}^{N} F_i(x) + \sum_{i=1}^{N}\left(\sum_{\ell=0}^{m(i)-1} F_N(T^{\ell N+i}x)\right) \\
&\qquad + \sum_{i=1}^{N} F_{k(i)}(T^{m(i)N+i}x) \\
&= \sum_{i=1}^{N} F_i(x) + \sum_{j=0}^{n-(i+k)} F_N(T^j x) + \sum_{i=1}^{N} F_{k(i)}(T^{m(i)N+i}x)
\end{aligned}
$$

(where in the last line we have tidied up the preceding expressions by appealing to the original definitions) and by dividing through by nN we get

$$
\begin{aligned}
\tfrac{1}{n}F_n(x) &\leq \frac{1}{N}\left(\frac{1}{n}\sum_{j=0}^{n-(i+k)} F_n(T^j x)\right) \\
&\quad + \frac{1}{n}\left(\frac{1}{N}\sum_{i=1}^{N} F_i(x) + \frac{1}{N}\sum_{i=1}^{N} F_{k(i)}(T^{m(i)N+i}x)\right) \qquad (2.13)
\end{aligned}
$$

We can deal with the right-hand side of (2.13) by observing that,

(a) for fixed N>0 and a.a.(m) $x \in$X, the expression $\frac{1}{N}\sum_{i=1}^{N} F_i(x)$ is obviously bounded, and using the Birkhoff ergodic theorem (Theorem 1.2) we can deduce the same of the term $\frac{1}{N}\sum_{i=1}^{N} F_{k(i)}(T^{m(i)N+i}x)$ and the difference

$$
\frac{1}{N}\sum_{j=n-(i+k)}^{n-1} F_N(T^j x) = \frac{1}{N}\left(\sum_{j=0}^{n-1} F_N(T^j x) - \sum_{j=0}^{n-(i+k)} F_N(T^j x)\right)
$$

Thus, for a.a.(m) $x \in$X,

$$
\frac{1}{n}\left(\frac{1}{N}\sum_{i=1}^{N} F_i(x) + \frac{1}{N}\sum_{i=1}^{N} F_{(n-i-m(i)N)}(T^{m(i)N+i}x) + \frac{1}{N}\sum_{j=n-(i+k)}^{n-1} F_N(T^j x)\right) \longrightarrow 0,
$$

as $n \longrightarrow +\infty$.

(b) applying the Birkhoff ergodic theorem we have (for N fixed)

$$\frac{1}{N}\Big(\frac{1}{n}\sum_{j=0}^{n-1} F_N(T^j x)\Big) \longrightarrow \frac{1}{N}\int F_N dm, \quad \text{as } n \longrightarrow +\infty.$$

In particular, letting $n \longrightarrow +\infty$ (with N fixed) in (2.13) gives,

$$F^+(x) \leq \frac{1}{N}\int F_N \, dm, \quad \forall N \geq 1 \text{ (for almost all } x) \tag{2.14}$$

(This should be viewed as a strengthening of (2.10).). Finally, we take the infinium over all $N>0$ to get

$$F^+(x) \leq \lambda = \inf\Big\{\frac{1}{N}\int F_N \, dm\Big\} \tag{2.15}$$

<u>*Step 3*</u>. In view of the inequality (2.15) it is clear that to complete the proof it suffices to show that $F^- \geq \lambda$. If $\lambda = -\infty$ we are finished, so henceforth we can assume that $\lambda > -\infty$.

A priori, we do not know that the function f^- is constant (as we did in the proof of Theorem 1.2). Therefore, we want to replace it by a more convenient function. We first fix $\epsilon>0$ and

(i) let $n(x) = \min\{\, n \geq 1 \mid \frac{1}{n} F_n(x) \leq f^-(x) + \epsilon\}$ (2.16)

(ii) let $A = \{x \in X \mid n(x) > N\}$, where N is chosen large enough that the set A satisfies

$$\int_A (\,|F_1(x)\,| + |F^-(x)\,|\,)\, dm\,(x) < \frac{\epsilon}{2} \tag{2.17}$$

As in the proof of Birkhoff's theorem we can introduce new functions

$$\widetilde{F}^-(x) = \begin{cases} F^-(x) & \text{if } x \notin A \\ F_1(x) & \text{if } x \in A \end{cases} \quad \text{and} \quad \widetilde{n}(x) = \begin{cases} n(x) & \text{if } x \notin A \\ 1 & \text{if } x \in A \end{cases} \tag{2.18}$$

By definition, the function $\widetilde{F}^-(x)$ satisfies $\widetilde{F}^-(x) = F^-(x)$ whenever $x \notin A$. Combining this with the estimate (2.17) when $x \in$A we get

$$\int \widetilde{F}\, dm \leq \int F^- dm + \epsilon \tag{2.19}$$

Step 4. We are now in a position to complete the proof. We begin with the simple inequality

$$F_{\tilde{n}(x)}(x) \leq \tilde{F}^{-}\ \tilde{n}(x) + \tilde{n}(x)\ \epsilon. \tag{2.20}$$

(NB. If $x \notin A$ then this follows from (2.16). Conversely, if $x \in A$ then it follows directly from the definitions.) For any integer $L > N$ we can write

$$F_L(x) \leq \sum_{j=0}^{L-1} \tilde{F}^{-}(T^j x) + L\epsilon + Nf^{-} + \sum_{j=L-N}^{L-1} |\, F_1(T^j x)\,| \tag{2.21}$$

(This derivation is the same as in Step 3 of proof of the Birkhoff ergodic theorem.) Therefore, by integrating (2.21) (recalling that m is T-invariant) and dividing through by L we have

$$\begin{aligned} \lambda \leq \frac{1}{L}\int F_L \mathrm{d}m &\leq \int \tilde{F}^{-}\mathrm{d}m + \epsilon + \frac{N F^{-}}{L} + \frac{N}{L}\int |F_1|\, \mathrm{d}m \\ &\leq \left(\int F^{-}\mathrm{d}m + \epsilon\right) + \epsilon + \frac{N F^{-}}{L} + \frac{N}{L}\int |F_1|\, \mathrm{d}m \end{aligned} \tag{2.22}$$

where we bound the first term on the right-hand side using (2.19). If we let $L \longrightarrow +\infty$ and $\epsilon \longrightarrow 0$ in (2.22) then we get

$$\lambda \leq F^{-} \tag{2.23}$$

To conclude, we see that (2.23) and (2.15) give $\lambda \leq F^{-} \leq F^{+} \leq \lambda$. Therefore, we deduce that the limit exists, and is equal to λ (for almost all x) . □

Notes

The subadditive ergodic theorem was originally proved by Kingman in 1963, although the proof we give is a more recent proof by Katznelson and Weiss [Ktn-Wei] (along the same lines as their proof of the Birkhoff ergodic theorem, presented in chapter 2). A nice summary of the statements and references to generalizations occurs in [Walters], pp.230-236.

The Oseledec theorem was proved in 1968, although again we

have presented an alternative approach taken from an article of Ruelle [**Ruelle$_1$**] (based in turn on a simplified proof of the Oseledec theorem due to Raghunathan). A nice account also appears as Appendix A in [**Margulis**].

Finally, there exist generalizations of these results to infinite dimensional manifolds [**Ruelle$_4$**].

Chapter 3

Entropy

Given a diffeomorphism $f: M \longrightarrow M$ we now understand that the asymptotic behavior of the iterates $f^n: M \longrightarrow M$ for large n, as viewed using an ergodic measure $m \in \mathcal{M}_{erg}$, is quantified by the finite set of Liapunov exponents $\lambda_1, \cdots, \lambda_d$ ($d = \dim M$) associated to m. However, although the theorem of Oseledec tells us of the existence of the Liapunov exponents it does not in itself provide us with an effective way of estimating them. In this chapter we will be concerned with estimating the Liapunov exponents in terms of other dynamical quantities, namely the measure theoretic and topological entropies.

3.1 Measure theoretic entropy.

We shall dive straight into the definition of measure theoretic entropy (despite its undoubted complexity) and attempt to unravel a little of its meaning afterwards.

Assume that $f: M \longrightarrow M$ is a C^1 diffeomorphism on a compact manifold. We want to define a map $h_{meas}: \mathcal{M}_{inv} \longrightarrow \mathbb{R}^+$ which associates to each f-invariant probability measure m a positive real number $h_{meas}(m) \geq 0$, as follows.

(a) If $d: M \times M \longrightarrow \mathbb{R}^+$ is the metric on M (derived from the Riemannian metric) then for any $n \geq 1$ define a new metric by

$$d_n(x,y) = \max_{0 \leq i \leq n} d(f^i x, f^i y).$$

(b) $\forall 0 < \delta < 1$, $n \geq 1$, $\epsilon > 0$ we call a finite set $K \subseteq M$ an *(n,ϵ,δ)-covering set* if the union $\bigcup_{x \in K} D_\epsilon(x; d_n)$ of the ϵ-balls $D_\epsilon(x; d_n) = \{y \in M \mid d_n(x,y) < \epsilon\}$ centered at points $x \in K$ has m-measure greater than δ.

(c) Let $N(n,\epsilon;\delta)$ be the *smallest* possible cardinality of a (n,ϵ,δ)-covering set (i.e. $N(n,\epsilon;\delta)=\min\{\text{card}K \mid K \text{ is a } (n,\epsilon;\delta)\text{-covering set}\}$.)

Notice that for fixed n,δ the map $\epsilon \mapsto N(n,\epsilon;\delta)$ is monotone decreasing and for fixed n,ϵ the map $\delta \mapsto N(n,\epsilon;\delta)$ is monotone increasing.

(d) Finally we define the *measure theoretic entropy* of the measure $m \in \mathcal{M}_{\text{inv}}$ by

$$h_{\text{meas}}(m) = \lim_{\delta\to 1} \lim_{\epsilon\to 0} \varliminf_{n\to+\infty} \frac{1}{n} \log N(n,\epsilon;\delta) \geq 0$$

When there is an confusion about the diffeomorphism f in question, we shall denote the measure theoretic entropy by $h_{\text{meas}}(m,f)$. It is not difficult to show that the entropy of any iterate f^n of f is given by $h_{\text{meas}}(m,f^n) = n\, h_{\text{meas}}(m,f)$, for any $n \in \mathbb{Z}$.

Remarks. (i) The definition above is due to Katok, but is equivalent to the classical (and more general) definition of Kolmogorov-Sinai in the case of diffeomorphisms [Katok]. We shall explain this further in Section 3.7. The usual Kolmogorov-Sinai definition works for any measurable space (*without* requiring a metric).

(ii) The definition is undoubtedly complicated. Intuitively, the use of the metrics d_n suggests that $h_{\text{meas}}(m)$ quantifies the *maximum asymptotic distortion of along orbits of* $f\colon M \longrightarrow M$ (on sets which are significant in terms of the invariant measure m) .

(iii) In fact, $\lim_{\epsilon\to 0} \varliminf_{n\to+\infty} \frac{1}{n}\log N(n,\epsilon;\delta)$ is *independent* of δ. Therefore, the final limit $(\delta \to 1)$ in the definition of $h_{\text{meas}}(m)$ is redundant (see [Katok]).

(iv) The function $h_{\text{meas}}\colon \mathcal{M}_{\text{inv}} \longrightarrow \mathbb{R}^+$ is affine on the convex set $\mathcal{M}_{\text{inv}}$ (i.e. $\forall\, m_1,m_2 \in \mathcal{M}_{\text{inv}}$, $0 \leq \alpha \leq 1$, $h_{\text{meas}}(\alpha m_1 + (1-\alpha)m_2) = \alpha\, h_{\text{meas}}(m_1) + (1-\alpha)\, h_{\text{meas}}(m_2)$) (see [Walters], p.183).

To give some appreciation of the meaning of the measure theoretic entropy it is useful to consider a few familiar examples.

Examples. (i) Let $M=\mathbb{R}^2/\mathbb{Z}^2$ and let $f\colon M \longrightarrow M$ be the diffeomorphism defined by $f(x_1,x_2)+\mathbb{Z}^2=(x_1+\alpha_1,x_2+\alpha_2)+\mathbb{Z}^2$ where $(\alpha_1,\alpha_2)\in\mathbb{R}^2$. As we have observed before, the Lebesgue-Haar measure m is an f-invariant ergodic measure. Since f is an isometry we see that $d(f^i x, f^i y) = d(x,y)$, for $i \geq 1$, and therefore $d_n(x,y) = d(x,y)$, for $n \geq 1$. This has the consequence for the above definitions that $N(n,\epsilon;\delta) \equiv N(1,\epsilon;\delta)$ for $n \geq 1$ and therefore

$$h_{\text{meas}}(m) = \lim_{\delta \to 1} \lim_{\epsilon \to 0} \left(\lim_{n \to +\infty} \frac{1}{n} \log N(1,\epsilon;\delta) \right) = 0.$$

(ii) Let $M = \mathbb{R}^2/\mathbb{Z}^2$ and let $f: M \to M$ be the diffeomorphism defined by $f(x_1,x_2) + \mathbb{Z}^2 = (2x_1 + x_2, x_1 + x_2) + \mathbb{Z}^2$. Since the origin (0,0) is a fixed point for f we know that the Dirac measure $m = \delta_{(0,0)}$ supported on (0,0) is invariant $m \in \mathcal{M}_{\text{inv}}$. By inspecting the above definition we see that the single element set $K = \{(0,0)\}$ is always a (n,ϵ,δ)-covering set for this measure. In particular, this means that $N(1,\epsilon;\delta) \equiv 1$ and consequently

$$h_{\text{meas}}(m) = \lim_{\delta \to 1} \lim_{\epsilon \to 0} \lim_{n \to +\infty} \left(\frac{1}{n} \log 1 \right) = 0.$$

(iii) We can consider the same manifold and diffeomorphism $f: M \longrightarrow M$ as in example (ii). However, this time we shall consider the invariant measure m to be the Lebesgue-Haar measure. As we have noted before, the tangent map of the diffeomorphism always takes the form of the matrix $Df = \left[\begin{smallmatrix} 2 & 1 \\ 1 & 1 \end{smallmatrix}\right]$ which has eigenvalues $\lambda_1, \lambda_2 = (3 \pm \sqrt{5})/2$. It is then easy to see that there exist constants c,C>0 such that for n sufficiently large

$$c \leq \frac{N(n,\epsilon;\delta)}{\lambda_1^n} \leq C$$

and therefore $h_{\text{meas}}(m) = \lim_{\delta \to 1} \lim_{\epsilon \to 0} \lim_{n \to +\infty} \frac{1}{n} \log(\lambda_1^n) = \log \lambda_1$.

<u>*A brief historical note*</u>. The measure theoretic entropy $h_{\text{meas}}(m)$ plays a *very* important role in the history of ergodic theory. It was originally introduced by Kolmogorov (and the definition further refined by Sinai) as an invariant for the classification of measure-preserving maps. Given two measure preserving transformations

$$\begin{cases} f_1: X_1 \longrightarrow X_1, & m_1 \in \mathcal{M}_{\text{inv}} \text{ (for } f_1) \\ f_2: X_2 \longrightarrow X_2, & m_2 \in \mathcal{M}_{\text{inv}} \text{ (for } f_2) \end{cases}$$

we call f_1 and f_2 *isomorphic* (and denote this by $f_1 \sim f_2$) if there exists a measure-preserving map $h: X_1 \longrightarrow X_2$ between the two spaces (i.e. $h^* m_1 = m_2$) such that the following diagram commutes

$$\begin{array}{ccc} X_1 & \xrightarrow{f_1} & X_1 \\ \downarrow h & & \downarrow h \\ X_2 & \xrightarrow{f_2} & X_2 \end{array}$$

The measure theoretic entropy was shown by Kolmogorov to satisfy the following:

Lemma 3.1 The measure theoretic entropy $h_{\text{meas}}(m)$ is an isomorphism invariant (i.e. $(f_1,m_1) \sim (f_2,m_2) \Rightarrow h_{\text{meas}}(m_1) = h_{\text{meas}}(m_2)$)

(see [Walters], p.89).

3.2 Measure theoretic entropy and Liapunov exponents.

We now return to the fundamental problem we mentioned at the beginning of this chapter:

Problem How can we estimate the Liapunov exponents $\lambda_1,\cdots,\lambda_d$?

One approach to solving this problem is to use the measure theoretic entropy. To proceed we need to be able to relate the measure theoretic entropy to the Liapunov exponents for a given ergodic measure $m \in \mathcal{M}_{\text{erg}}$.

Theorem 3.1 (Pesin-Ruelle inequality). Given a C^1 diffeomorphism $f: M \longrightarrow M$ and an ergodic measure $m \in \mathcal{M}_{\text{erg}}$ with associated Liapunov exponents $\lambda_1,\cdots,\lambda_d$ we have the following inequality

$$\left(\sum_{\lambda_i > 0} \lambda_i\right) \geq h_{\text{meas}}(m). \tag{3.1}$$

The proof of Theorem 3.1 will be postponed to the end of this chapter.

Remarks (i) If we replace the diffeomorphism f by its inverse f^{-1} then this diffeomorphism has the same ergodic measure m with the same entropy but the change has the effect of changing the signs of the Liapunov exponents i.e. $\lambda_i \longrightarrow -\lambda_i$. If we now apply Theorem 3.1 to the diffeomorphism f^{-1} then (3.1) takes the form

$$-h_{\text{meas}}(m) \geq \Big(\sum_{\lambda_i<0} \lambda_i \Big).$$

(ii) It is very natural to ask when there is an *equality* in (3.1). When m is a smooth measure (i.e. equivalent to the volume on m) and $f: M \longrightarrow M$ is C^2 then Pesin showed that there is actually an equality (see [Mañé$_2$], p.342). More recently, necessary and sufficient conditions on m for an equality in (3.1), again assuming f is a C^2 diffeomorphism, have been given by Ledrappier and Young [Led-You]. We shall return to this subject again in the final chapter. However, by contrast there exist examples of C^∞ diffeomorphisms $f: M \longrightarrow M$ for which equality in (3.1) is never realised for *any* ergodic measure $m \in \mathcal{M}_{\text{erg}}$. (In view of Pesin's result above we see that these examples must *not* have smooth invariant measures.) In the standard example M is the 2-sphere S^2 and f is the diffeomorphism described in Figure 10.

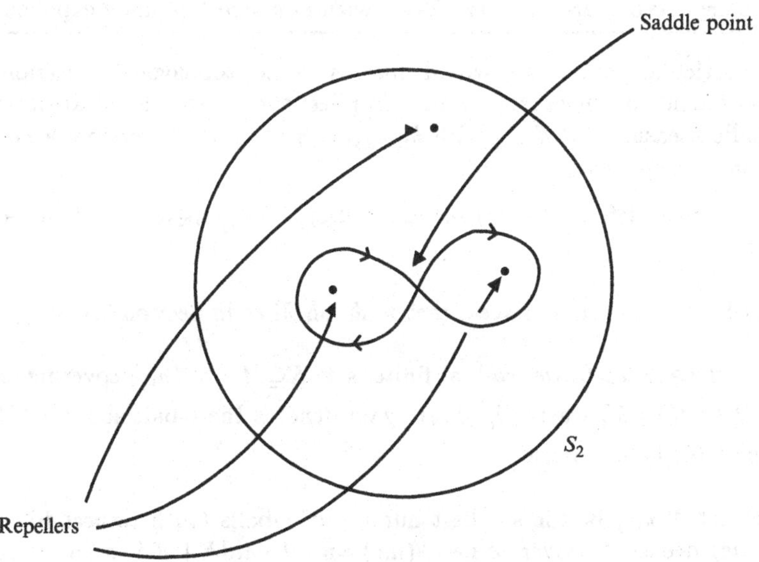

Figure 10: A diffeomorphism with only strict inequality in (3.1)

In this example all invariant measures in $\mathcal{M}_{\text{inv}}$ must be supported on the four fixed points indicated. Since these points are either repellers or the saddle points there are always strictly positive Liapunov exponents, irrespective of the invariant measure. However, any measure with finite support necessarily satisfies $h_{\mathrm{meas}}(m)=0$, thus we *never* have an equality in (3.1) for this example.

3.3 Topological entropy.

Theorem 3.1 gives us a criterion for the existence of non-zero Liapunov exponents for an ergodic measure $m\in\mathcal{M}_{\mathrm{erg}}$ associated to a C^1 diffeomorphism $f\colon M\longrightarrow M$. In particular, we see that it suffices to show that $h_{\mathrm{meas}}(m)>0$. However, this does not immediately seem to be much of an improvement since in general the calculation of the measure theoretic entropy from its definition does not seem to be any more tractable than the computation of the Liapunov exponents.

With this point in mind we can pose the following problem:

<u>*Problem*</u>. When are there $m\in\mathcal{M}_{\mathrm{erg}}$ with non-zero Liapunov exponents?

In particular, what we would like is some topological criterion (i.e. independent of measures) which implies the existence of at least one ergodic measure $m\in\mathcal{M}_{\mathrm{erg}}$ with $h_{\mathrm{meas}}(m,f)>0$, and therefore a non-zero Liapunov exponent.

We define the *topological entropy* $h_{\mathrm{top}}(f)\in\mathbb{R}^+$ of $f\colon M\longrightarrow M$ as follows.

(a) For each $n\geq1$ we have a metric d_n on M as in Section 3.1.

(b) $\forall n\geq1$, $\epsilon>0$ we call a finite set $K\subseteq M$ an (n,ϵ)-covering set if $\bigcup_{x\in K} D_\epsilon(x;d_n)=M$, where $D_\epsilon(x;d_n)$ again denotes the ϵ-ball about $x\in M$ with respect to the d_n-metric.

(c) Let $N(n,\epsilon)$ be the smallest number of ϵ-balls (with respect to the d_n-metric) needed to cover M i.e. $N(n,\epsilon)=\min\{\ \mathrm{card}K \mid K \text{ is a } (n,\epsilon)\text{-covering set}\}$.

Notice that for each $n\geq1$, $\epsilon\mapsto N(n,\epsilon)$ is monotone decreasing.

(d) Define the *topological entropy* by

$$h_{\mathrm{top}}(f) = \lim_{\epsilon\to 0} \varlimsup_{n\to+\infty} \frac{1}{n} \log N(n,\epsilon) \geq 0$$

Remarks. (i) The definition actually makes sense for any continuous map on a compact metric space.

(ii) If $N \subseteq M$ is an invariant sub-manifold (i.e. $fN \subseteq N$), then it is easy to see from the definitions that $h_{\mathrm{top}}(f|N) \leq h_{\mathrm{top}}(f|M)$.

(iii) The intuitive interpretation of the topological entropy $h_{\mathrm{top}}(f)$ is that it measures the *asymptotic distortion of the iterates* f^i*:* $M \longrightarrow M$ *along orbits.*

The best way to understand this definition is, again, to consider our standard examples.

Examples. (i) Let $M = \mathbb{R}^2/\mathbb{Z}^2$ and define $f: M \longrightarrow M$ by $f(x_1,x_2)+\mathbb{Z}^2 = (x_1+\alpha_1, x_2+\alpha_2)+\mathbb{Z}^2$. Since f is an isometry and $d_n(x,y) = d(x,y)$, for all $n \geq 1$, it is easy to see that $N(n,\epsilon) \equiv N(1,\epsilon)$. In particular, if we substitute directly into the definition we see that

$$h_{\mathrm{top}}(f) = \lim_{\epsilon\to 0}\left(\varlimsup_{n\to+\infty} \frac{1}{n} \log N(n,\epsilon)\right) = 0.$$

(ii) Let $M = \mathbb{R}^2/\mathbb{Z}^2$ and let $f: M \longrightarrow M$ by $f(x_1,x_2)+\mathbb{Z}^2 = (2x_1+x_2, x_1+x_2)+\mathbb{Z}^2$. By considering the eigenvalues of the matrix representing the constant matrix Df we can bound the quantity $N(n,\epsilon)$, for sufficiently large n, in terms of the nth power λ_1^n of the largest eigenvalue $\lambda_1 = (3+\sqrt{5})/2$. In particular, we can calculate

$$h_{\mathrm{top}}(f) = \lim_{\epsilon\to 0} \left(\varlimsup_{n\to+\infty} \frac{1}{n} \log N(n,\epsilon)\right) = \log \lambda_1.$$

A brief historical note. Originally $h_{\mathrm{top}}(f)$ was introduced as an invariant for the classification of continuous maps i.e. given two homeomorphisms $f_2: X_2 \to X_2$ and $f_1: X_1 \to X_1$ of compact topological spaces, we call f_1, f_2 *(topologically) conjugate* (and denote this by $f_1 \approx f_2$) if there exists a homeomorphism $h: X_1 \longrightarrow X_2$ such that the following diagram commutes

$$\begin{array}{ccc} X_1 & \xrightarrow{f_1} & X_1 \\ \downarrow h & & \downarrow h \\ X_2 & \xrightarrow[f_2]{} & X_2 \end{array}$$

The importance of $h_{top}(f)$ for classifying homeomorphisms can be seen by the following result.

Lemma 3.2 $h_{top}(f)$ is a conjugacy invariant (i.e. $f_1 \approx f_2 \Rightarrow h_{top}(f_1) = h_{top}(f_2)$)

(see [Walters], p.167).

Our motivation for introducing $h_{top}(f)$ was to find a way to show the existence of measures $m \in \mathcal{M}_{erg}$ with $h_{meas}(m) > 0$. The relationship between the topological entropy $h_{top}(f)$ of the map $f: M \longrightarrow M$ and the measure theoretic entropy $h_{meas}(m)$ of an ergodic measure $m \in \mathcal{M}_{inv}$ is given by the following simply stated result.

Proposition 3.1 (Variational principle).

(i) $h_{meas}(m) \leq h_{top}(f), \ \forall \ m \in \mathcal{M}_{inv}$.

(ii) $h_{top}(f) = \sup \{ h_{meas}(m) \ | \ m \in \mathcal{M}_{inv} \}$.

This result is now a well-established part of ergodic theory. Part (i) is immediate from our choice of definitions. We omit the proof of part (ii) and refer the reader to [Mañé$_2$], p.316 or [Walters], p.187 for excellent accounts.

Remark. In all our examples we were able to find $m \in \mathcal{M}_{inv}$ which attains the supremum in part (ii). However, we mention for completeness that there exist examples for which no measure satisfies $h_{meas}(m) = h_{top}(f)$ (cf.[Walters],p.193, and we shall demonstrate such an example in 6.7).

3.4. Topological entropy and Liapunov exponents.

We want to give a criterion on the topological entropy of $f: M \longrightarrow M$ which guarantees the existence of ergodic measures $m \in \mathcal{M}_{erg}$ with at least one non-zero Liapunov exponent (and in the case of surfaces gives that *both* Liapunov exponents are non-zero).

Theorem 3.2. Let $f: M \longrightarrow M$ be a C^1 diffeomorphism with $h_{top}(f) > 0$, then

(i) $\exists\ m \in \mathcal{M}_{erg}$ with a non-zero Liapunov exponent,

(ii) if $\dim M = 2$ then $\exists m \in \mathcal{M}_{erg}$ with Liapunov exponents $\lambda_1 > 0 > \lambda_2$.

Proof. We begin with an observation common to the proof of both parts. By part (ii) of the variational principle (Proposition 3.1) we can choose an invariant measure $m \in \mathcal{M}_{inv}$ with $h_{top}(f) \geq h_{meas}(m) > 0$. Without loss of generality we can assume that the measure m is ergodic since,

(a) $\text{Ext}(\mathcal{M}_{inv}) = \mathcal{M}_{erg}$ (by Lemma 1.4), and

(b) $m \mapsto h_{meas}(m)$ is an affine map on the convex space $\mathcal{M}_{inv}$ (cf. Remark (iv) in Section 3.1), and we can therefore replace m (if necessary) by an ergodic measure $m' \in \mathcal{M}_{erg}$ which occurs in the ergodic decomposition of m and still assume $h_{meas}(m') > 0$.

To prove part (i) of the theorem, we observe from Theorem 3.1 that

$$\Big(\sum_{\lambda_i > 0} \lambda_i\Big) \geq h_{meas}(m) > 0,$$

and thus, in particular, $\exists \lambda_i > 0$.

To prove part (ii) of the theorem, we note that in the case of surfaces there are just two Liapunov exponents. Applying Theorem 3.1 when there are only two Liapunov exponents gives

$$0 < h_{meas}(m) \leq \max\{0, \lambda_1\} + \max\{0, \lambda_2\} \tag{3.2}$$

If we replace the diffeomorphism f by its inverse f^{-1} and use the same ergodic measure m then the Liapunov exponents merely change sign (i.e. $\lambda_i \longrightarrow -\lambda_i$, i=1,2), but the measure theoretic entropy remains the same (see Section 3.1, Remark (i)). Thus (3.2) is replaced by

$$0 < h_{meas}(m) \leq \max\{0, -\lambda_1\} + \max\{0, -\lambda_2\} \tag{3.3}$$

However, (3.2) and (3.3) can only be satisfied when either $\lambda_1>0>\lambda_2$ or $\lambda_2>0>\lambda_1$. □

Remark. If we drop the assumption that M is a surface in part (ii) of Theorem 3.2 and allow $\dim M \geq 3$ then it is no longer necessarily true that $h_{top}(f)>0$ implies the existence of an ergodic measure $m\in\mathcal{M}_{erg}$ with no zero Liapunov exponents. (An easy counter-example is the discrete map corresponding to the 'time one' geodesic flow for a compact negatively curved surface where every invariant measure must have a zero Liapunov exponent corresponding to the flow direction.)

Now we return to the important point of finding a *useful* criterion for the existence of ergodic measures $m\in\mathcal{M}_{erg}$ with non-zero Liapunov exponents. Theorem 3.2 above shows that this is true if $h_{top}(f)>0$. Thus we have reduced the problem to the following

Problem. When does $f: M\longrightarrow M$ satisfy $h_{top}(f) > 0$?

The definition of $h_{top}(f)$ is again too complicated to make explicit computation a realistic approach in the general case. However, in some instances there exist very simple topological criteria that guarantee $h_{top}(f)>0$. For the case $\dim M=2$ we have the following examples.

(a) Let $f^*: H_1(M,\mathbb{R})\longrightarrow H_1(M,\mathbb{R})$ *denote the induced action of* $f: M\longrightarrow M$ on the first (real) homology group of the surface M. This action will be represented by a matrix (with integer entries). If f^* has eigenvalues outside the unit circle $K = \{z\in\mathbb{C}|\ |z|=1\}$ then $h_{top}(f)>0$ (after Manning, see [Mañé$_2$], p.382). This is a special case of the Shub entropy conjecture, the discussion of which we shall return to in Section 6.6 when we relate Yomdin's contribution to the problem.

(b) Assume that M is an orientable surface of genus g and that $f: M\longrightarrow M$ is an orientation-reversing map. Furthermore, we assume that there are *at least* $(g+2)$ periodic points with distinct odd periods. Then $f: M\longrightarrow M$ satisfies $h_{top}(f)>0$; see [Handel]. (The proof of the theorem uses a rather ingenious application of Thurston's work on the isotopy classes of surface homeomorphisms.)

3.5 Equivalent definitions of measure theoretic entropy.

The definition of measure theoretic entropy, for a diffeomorphism $f: M \longrightarrow M$ and an invariant measure $m \in \mathcal{M}_{\text{inv}}$, which we gave earlier in this chapter turns out to be well suited for applications, as we shall see later. However, it does differ from the original (and more familiar) definition of Kolmogorov and Sinai. We begin by recalling this more standard definition.

(a) Let $\alpha = \{A_1, \cdots, A_n\}$ be any finite measurable partition for $(M, \mathcal{B}, m)$ (i.e. $m\left(M - \bigcup_{i=1}^{n} A_i\right) = 0$ and $m(A_i \cap A_j) = 0$, for $i \neq j$). For any $N \geq 1$ we can define another partition

$$\alpha^{(N)} = \{A_{i_0} \cap f^{-1} A_{i_1} \cap \cdots \cap f^{-(N-1)} A_{i_{N-1}} \mid 1 \leq i_0, \cdots i_{N-1} \leq n\}.$$

(b) We can introduce $H(\alpha^{(N)}) = -\sum_{C \in \alpha^{(N)}} m(C) \log m(C) \geq 0.$

(c) We define the *entropy* of a partition α by $h(m, \alpha) = \overline{\lim_{N \to \infty}} \frac{1}{n} H(\alpha^{(N)}) \geq 0$ (in fact the limit actually exists because $N \longrightarrow H(\alpha^{(N)})$ is a subadditive sequence; see [Bowen], p.30).

(d) Finally, we define the (Kolmogorov-Sinai) measure theoretic entropy of the measure $m \in \mathcal{M}_{\text{inv}}$ by

$$h_{\text{KS}}(m) = \sup\{h(m, \alpha) \mid \alpha = \text{finite measurable partition}\}$$

(see [Walters], §4.4 for details).

One of the nice things about this definition is that frequently we can avoid taking the supremum over all partitions, in part (d), if we make a good choices of partition α. More precisely

Lemma 3.5

(i) $h(m, \alpha_i) \longrightarrow h_{\text{KS}}(m)$ if $\alpha_j \longrightarrow \mathcal{B}$, as $j \longrightarrow +\infty$ (i.e. $\forall \epsilon > 0$ and $B \in \mathcal{B}$, $\exists i \geq 1$ and $C_1, \cdots, C_r \in \alpha_i$ with $m\left(B \,\Delta \bigcup_{i=1}^{r} C_i\right) < \epsilon$).

(ii) $h_{\text{KS}}(m) = h(m,\alpha)$ if α is a generating partition (i.e. $\forall\epsilon>0$ and $B\in\mathfrak{B}$, $\exists N\geq 1$ and $C_1,\cdots,C_r\in\alpha^{(N)}$ with $m\left(B \,\Delta \bigcup_{i=1}^{r} C_i\right)< \epsilon$.

In the above lemma Δ denotes the symmetric difference; see Appendix A. These are both standard results in ergodic theory. Part (i) can be found in [Parry], p.61, for example. Part (ii) is a famous result to Kolmogorov and Sinai (see [Parry], p.62, or [Walters], p.95).

Remark (Kolmogorov-Sinai entropy and condition expectation). There is another simple formulation of the Kolomogorov-Sinai entropy which is closely related to the conditional expectation mentioned in Section 1.7. In fact,

$$h_{KS}(m) = \lim_{k\to\infty} \int I_m(\alpha| \bigvee_{i=1}^{k} f^{-i}\alpha)(x) \; dm(x)$$

where $I_m(\alpha|\beta)(x) = -\sum_{A\in\alpha} E_m(\chi_A|\beta)\log E_m(\chi_A|\beta)$, with $E_m(.|\beta)(x)$ denoting the conditional expectation function. The equivalence of the two definitions of $h(m,\alpha)$ is an easy exercise (see. [Parry], p.34, or [Walters], §4.3, for more details).

We want to show that the two definitions of entropy coincide for diffeomorphisms (and, in fact, more generally for homeomorphisms of compact metric spaces). In the course of the proof of this equality, the following standard result will prove useful.

Theorem 3.3 (Shannon). Let α be a generating partition and for $N\geq 1$ let $A^{(N)}(x)\in\alpha^{(N)}$ denote the element of the partition $\alpha^{(N)}$ containing $x\in M$ a.e.(m), i.e. $x\in A^{(N)}(x)$. Then

$$\lim_{N\to\infty} -\frac{\log m(A^{(N)}(x))}{N} = h_{\text{meas}}(m) \text{ , for a.a.}(m) \;\; x\in M$$

(see [Parry], p.39-41, especially Exercise 7).

This brings us to the promised result equating the two definitions of entropy.

Theorem 3.4 (Katok). If $f: M\longrightarrow M$ is a diffeomorphism of a compact manifold and $m\in\mathcal{M}_{\text{inv}}$ then $h_{\text{KS}}(m) = h_{\text{meas}}(m)$.

Proof. (The proof we give is not Katok's original proof, but one worked out in **[Mendoza]**, based on Misiurewicz's proof of the variational principle; cf. **[Katok]**.)

We begin by establishing the following inequality:

Claim 1. $h_{\text{meas}}(m) = \lim_{\delta\to 0} \lim_{\epsilon\to 0} \lim_{n\to 0} \frac{1}{n} \log N(n,\epsilon;\delta) \geq h_{\text{KS}}(m)$
(NB. This is the harder inequality.)

Step 1. Let $\alpha = \{A_1,\cdots,A_n\}$ be any measurable partition and let $\eta>0$ be some small value. We can always choose pairwise disjoint *compact sets* $B_i \subset A_i$ such that, if we add $B_0 = M - \bigcup_{i=1}^{n} B_i$, then the partition $\beta = \{B_0,B_1,\cdots,B_n\}$ is 'close to α in entropy', in the sense that $|h(m,\alpha)-h(m,\beta)|<\eta$ (the details only require a little more familiarity with the definitions of entropy; cf. **[Walters]**, pp.187-190).

Step 2. For $N\geq 1$ we define

$$Y_N = \{\, y\in M \mid -\frac{\log m(B^{(N)}(y))}{n} \geq h_{\text{KS}}(m)-\eta,\ \forall\ n\geq N\,\}$$

and by Shannon's theorem (Theorem 3.6) we have that $m(Y_N)\longrightarrow 1$ as $N\longrightarrow\infty$. Choose $\epsilon < \frac{1}{2}\min\{d(B_i,B_j) \mid i\neq j,\ 1\leq i,j\leq n\}$ then each set $D_\epsilon(x;d_N)\cap Y_N$ intersects at most 2^N elements of $\beta^{(N)}$. In particular, we can conclude that

$$m\Big(D_\epsilon(x;d_N)\cap Y_N\Big) \leq 2^N e^{-N(h_{\text{KS}}(m)\ -\eta)}. \tag{3.4}$$

Step 3. Given $1>\delta>0$, let $K\subset M$ be an $(n,\epsilon;\delta)$-covering set. We can assume that N is sufficiently large that $m(Y_N)>\delta$ and assume $n\geq N$. We can introduce the subset $K' = \{x\in K \mid D_\epsilon(x;d_n)\cap Y_N\neq\emptyset\}$. We want the following inequality on the cardinality of the finite set K.

$$\text{Card}(K) \geq C\, e^{n(h_{\text{KS}}(m)\ -\eta\ -\ \log 2)} \tag{3.5}$$

for some constant $C>0$. To see this we observe that

$$
\begin{aligned}
\operatorname{Card}(K) e^{-n(h_{KS}(m) - \eta - \log 2)} \\
&\geq \operatorname{Card}(K').e^{-n(h_{KS}(m) - \eta - \log 2)} && (\text{since } K' \subset K) \\
&\geq \sum_{x \in K'} m\Big(D_N(x;\epsilon) \cap Y_N\Big) && (\text{by } (3.4)) \\
&\geq m\Big(\Big(\bigcup_{x \in K'} D_N(x;\epsilon)\Big) \cap Y_N\Big) \geq m\Big(\Big(\bigcup_{x \in K} D_N(x;\epsilon)\Big) \cap Y_N\Big)
\end{aligned}
$$

(since $D_N(x;\epsilon) \cap Y_N \neq \emptyset$ for $x \in K - K'$)

$$\geq 2\delta - 1$$

since $m\Big(\bigcup_{x \in K} D_N(x;\epsilon)\Big) > \delta$ and $m(Y_N) > \delta$.

In particular, this lower bound is *independent* of n.

<u>Step 4</u>. Since (3.5) applies for all $(n,\epsilon;\delta)$-covering sets, it certainly applies for the one with the *smallest* cardinality (i.e. where $\operatorname{Card}(K) = N(n,\epsilon;\delta)$) and we can deduce that:

$$h_{\text{meas}}(m) = \lim_{\delta \to 1} \lim_{\epsilon \to 0} \lim_{n \to 0} \frac{1}{n} \log N(n,\epsilon;\delta) \geq h_{KS}(m) - \eta - \log 2 \quad (3.6)$$

We can improve the lower bound (3.6) by two simple steps.

(a) Since $\eta > 0$ can be chosen arbitrarily small, we can set it to zero in (3.6) to get $h_{\text{meas}}(m) \geq h_{KS}(m) - \log 2$.

(b) If we repeat the argument with $f: M \longrightarrow M$ replaced by its rth iterate $f^r: M \longrightarrow M$ $(r \geq 1)$ then, by standard properties of entropy, $h_{KS}(m)$ and $h_{\text{meas}}(m)$ are replaced by $r h_{KS}(m)$ and $r h_{\text{meas}}(m)$, respectively.

Dividing out by a factor of r we get

$$h_{\text{meas}}(m) \geq h_{KS}(m) - \frac{\log 2}{r}$$

and letting $r \longrightarrow +\infty$ gives $h_{\text{meas}}(m) \geq h_{KS}(m)$, completing our proof of Claim 1.

To complete the proof of the theorem, we need to establish the reverse inequality.

Claim 2: $h_{\text{meas}}(m) = \lim_{\delta \to 1} \lim_{\epsilon \to 0} \lim_{n \to 0} \frac{1}{n} N(n,\epsilon;\delta) \leq h_{\text{KS}}(m)$
(NB. This is the easier inequality.)

Given $\epsilon>0$, choose a measurable finite partition $\alpha = \{A_1,\cdots,A_n\}$ with $\text{diam}(A_i)\leq\epsilon$. Given $1>\delta>0$, and any small value $\eta>0$, then for $N\geq 1$ sufficiently large the set

$$X_N = \left\{ x\in M \mid -\frac{\log m(A^{(N)}(x))}{n} \leq h_{\text{KS}}(m) + \eta, \ \forall\ n\geq N \right\}$$

will satisfy $m(X_N)>\delta$ (by Shannon's theorem; Theorem 3.6). In particular, we can choose points $K=\{x_1,\cdots,x_\ell\}\subset X_N$ for which the sets $C_i=A^{(N)}(x_i)\in\alpha^{(N)}$, i=1,$\cdots$,$\ell$, are *distinct*, and such that:

(a) $C_i \subset D_\epsilon(x_i,d_N)$;

(b) $X_N \subset \bigcup_{i=1}^{m} C_i$; and

(c) $m(C_i) \leq e^{-N(h_{\text{KS}}(m)+\eta)}$.

In particular, we see from (a) and (b) that $X_N \subset \bigcup_{x\in K} D_\epsilon(x;d_N)$ and conclude that

$$m\Big(\bigcup_{x\in K} D_\epsilon(x;d_N)\Big)\geq m(x_N)>\delta$$

i.e. that K is a $(n,\epsilon;\delta)$-covering set. From (c) we see that

$$1 \geq m\Big(\bigcup_{i=1}^{\ell} C_i\Big) = \sum_{i=1}^{\ell} m(C_i) \geq \ell.e^{-N(h_{\text{KS}}(m)+\eta)}$$

and so we deduce that $\text{Card}(K) = \ell \leq e^{N(h_{\text{KS}}(m)+\eta)}$. In particular, since $\text{Card}(K) \geq N(n,\epsilon;\delta)$ we see that

$$h_{\text{meas}}(m) = \lim_{\delta\to 1} \lim_{\epsilon\to 0} \lim_{n\to 0} \frac{1}{n} \log N(n,\epsilon;\delta) \leq h_{\text{KS}}(m) + \eta$$

Finally, since $\eta>0$ can be chosen arbitrarily small we see that $h_{\text{meas}}(m) \leq h_{\text{KS}}(m)$, which completes the proof of the reverse inequality.

Comparing the inequalities in different directions in Claims 1 and 2 we get the required equality, and this completes the proof of the theorem. □

3.6. Proof of Pesin-Ruelle inequality (Theorem 3.1). We shall follow the original proof in [Ruelle$_2$] (where Ruelle explains that his proof was developed from a part of the work of Pesin in a more specific context).

We shall assume, for simplicity, that M is a surface, i.e. $\dim M=2$. The general case is very similar, but we benefit from some simplifications in the notation. We shall divide the proof into its principal component steps.

Step 1. We want to divide the surface into small squares. We first 'trivialize' the surface by choosing a smooth triangulation of M into simplices (see Appendix C). Using the natural co-ordinates of the associated simplex in the plane

$$\Delta = \{(x,y) \mid 0\leq x, y\leq 1, x+y\leq 1\}$$

and given any $\epsilon>0$, we can next sub-divide each triangle into ϵ-squares $\{(x,y) \mid r\epsilon\leq x\leq(r+1)\epsilon, s\epsilon\leq x\leq(s+1)\epsilon\}$, $r,s\in\mathbb{N}$ (or ϵ-small triangles next to the boundary $x+y=1$).

We can assume without loss of generality that the boundaries of the cubes have zero measure with respect to a given invariant measure m. (If this were not the case then the triangulation need only be moved an arbitrarily small distance for it to be true.) We denote this partition as $\mathcal{P}_\epsilon$.

Lemma 3.3 $\exists C>0$, $\forall n\geq 1$, $\exists \epsilon(n)>0$ such that $\forall 0<\epsilon<\epsilon(n)$, $A\in\mathcal{P}_\epsilon$, $\mathrm{Card}\{B\in\mathcal{P}_\epsilon \mid A\cap f^n(B)\neq\emptyset\} \leq C\max\{\|D_xf^n\| + |\mathrm{Det}(D_xf^n)| \mid x\in B\}$.

Proof of Lemma 3.3. For sufficiently small $\epsilon>0$, we can linearize the derivative D_xf^n (for a fixed $n\geq 1$) in an ϵ-neighborhood of x. The estimate on $\|D_xf\|$ gives a bound on the (one-dimensional) diameter of the image $f^n(A)$. The estimate on $|\mathrm{Det}(D_xf^n)|$ gives a bound on the (two-dimensional) volume of the image $f^n(A)$. Since the majority of the elements $B\in\mathcal{P}$ are of comparable size, this provides an upper bound on the cardinality above. □

Step 2. We need an upper bound for the entropy $h(m,\mathcal{P}_\epsilon)$ of the partition $\mathcal{P}_\epsilon$, with respect to diffeomorphism $g=f^N$.

Lemma 3.4 $\forall n \geq 1,\ 0 < \epsilon < \epsilon(n)$

$$h(m,\mathcal{P}_\epsilon) \leq \log C + \int \max\{\log \|D_x f^n\|,\ |\mathrm{Det}(D_x f^n)|\}dm. \quad (3.7)$$

Proof of Lemma 3.4. We prefer the alternative definition of $h(m,\mathcal{P}_\epsilon)$ given in Section 3.5:

$$h(m,\mathcal{P}_\epsilon) = \lim_{k\to\infty} \int I_m(\mathcal{P}_\epsilon | \bigvee_{i=1}^{k} g^{-i}\mathcal{P}_\epsilon)(x)\ dm(x)$$

with

$$I_m(\mathcal{P}_{\epsilon(n)} | \bigvee_{i=1}^{k} g^{-i}\mathcal{P}_\epsilon)(x)$$

$$= -\sum_{A\in\mathcal{P}_\epsilon} E_m(\chi_A | \bigvee_{i=1}^{k} g^{-i}\mathcal{P}_\epsilon)\log E_m(\chi_A | \bigvee_{i=1}^{k} g^{-i}\mathcal{P}_\epsilon)$$

$$= -\sum_{A\in\mathcal{P}_\epsilon} \left(\sum_{B\in \bigvee_{i=1}^{k} g^{-i}\mathcal{P}_\epsilon} \frac{m(A\cap B)}{m(B)} \log \frac{m(A\cap B)}{m(B)}\ \chi_B(x) \right) \quad (3.8)$$

Observe that $x \mapsto -x\log(x)$ is bounded by $\frac{1}{e}$; then we see that for each set $B = f^{-1}A_1 \cap \ldots \cap f^{-k}A_k \in \bigvee_{i=1}^{k} f^{-i}\mathcal{P}_\epsilon$ (where $A_1,\ldots,A_k \in \mathcal{P}_\epsilon$), and each point $x\in\mathcal{P}_\epsilon$, we can bound (3.8) as follows

$$\begin{aligned} I_m(\mathcal{P}_\epsilon | \bigvee_{i=1}^{k} g^{-i}\mathcal{P}_\epsilon)(x) &\leq \mathrm{Card}\{A\in\mathcal{P}_\epsilon |\ A\cap B \neq \emptyset\} \\ &\leq \mathrm{Card}\{A\in\mathcal{P}_\epsilon \mid A\cap g(A_1) \neq \emptyset\} \\ &\leq C\max\{\|D_x f^n\| + |\mathrm{Det}(D_x f^n)| \mid x\in A_1\} \end{aligned}$$

(using Lemma 3.3).

Letting $k \longrightarrow +\infty$ completes the proof of Lemma 3.4. □

Step 3. We want to fix $n \geq 1$ and let $\epsilon \longrightarrow 0$. Since $\mathcal{P}_\epsilon$ approximates the Borel σ-algebra, $h(m,\mathcal{P}_\epsilon) \longrightarrow h_{\mathrm{meas}}(m,f^n)$, by Lemma 3.5 (ii). Thus (3.7) gives

$$h_{\mathrm{meas}}(m,f^n) \leq \log C + \int \max\{\log \|D_x f^n\|, \log |\mathrm{Det}(D_x f^n)|\}dm \quad (3.9)$$

Observe that $\frac{1}{n}\log\|D_x f^n\| \longrightarrow \lambda_1$, as $n \longrightarrow +\infty$, for a.e.$(m)$ $x \in M$, by the Oseledec theorem (Theorem 2.1). Similarly, the subadditive ergodic theorem (Proposition 2.2) applied to $F_n(x) = \log|\mathrm{Det}(D_x f^n)|$ gives that

$$\frac{1}{n}\log|\mathrm{Det}(D_x f^n)| \longrightarrow \lambda_1 + \lambda_2, \text{ as } n \longrightarrow +\infty, \text{ for a.e.}(m) \ x \in M.$$

Since $h_{\text{meas}}(m, f^n) = n h_{\text{meas}}(m, f)$ and C is *independent* of $n>0$, we can divide both sides of (3.9) by n and let $n \longrightarrow +\infty$ to get the final inequality $h_{\text{meas}}(m,f) \leq + \left(\sum_{\lambda_i > 0} \lambda_i\right)$. This completes the proof of Theorem 3.1. □

3.7 Oseledec's theorem, topological entropy and Lie theory.

Oseledec's theorem was found by Margulis to have important applications to the theory of Lie groups of higher rank. Following Mostow's rigidity theorem for Lie groups of rank greater than or equal to 2, Margulis proved a stronger 'arithmeticity rigidity' result for arithmetic groups, which required Oseledec's theorem (although, it is apparently no longer a necessary ingredient in more modern proofs).

A brief account of the development of the theory is contained in the first few pages of the introduction to Zimmer's book [Zimmer], and a very complete introduction to this area is contained in Margulis' book [Margulis].

To give just a little flavor of the type of results involved we shall consider the statement of just one result. Let $G = SL(n, \mathbb{R})$, with $n \geq 2$, and let M be a compact C^2 Riemannian manifold. Let $A\colon G \times M \longrightarrow M$ be a C^2 action (i.e. a smooth map satisfying $A(g_2, A(g_1, x)) = A(g_2 g_1, x)$, $\forall x \in M$, $\forall g_1, g_2 \in G$. We assume, in addition, that:

(i) the action preserves some smooth (i.e. equivalent to the volume) measure m (i.e. $m(A(g,B)) = m(B)$, $\forall g \in G$);

(ii) the action is ergodic (i.e. $A(g,B) = B, \forall g \in G \Rightarrow$ either $m(B)=0$ or $m(M-B)=0$).

We can define a C^2 diffeomorphism $f\colon M \longrightarrow M$ by fixing an element $g \in G$ and restricting the action so that $f(x) = A(g,x)$, $\forall x \in M$.

Theorem 3.5 (Margulis). *Either* the topological entropy $h(f)=0$, or

$$h(f) = \log\Big(\max\{|\rho| \mid \rho \text{ is an eigenvalue for } \xi(g)\}\Big)$$

for some representation ξ: $SL(n,\mathbb{R})\to SL(m,\mathbb{C})$, $m\in\mathbb{N}$.

In particular, this theorem has the following very striking corollary.

Corollary 3.5.1 (Entropy rigidity). If we fix an element $g\in G$, but let the C^2 action A: $G\times M\longrightarrow M$ vary, the topological entropy $h(f)$ of the associated diffeomorphism $A(g,.)=f$: $M\longrightarrow M$ can take only finitely many values; see [Furstenberg].

Remark. Let G be a semi-simple Lie group; then it can be written in the form $G=KAN$, where $A\subset G$ is a maximal abelian subgroup, $K\subset G$ a maximal compact group, and N nilpotent. Assume that M is a compact locally symmetric space formed from G, then there is a natural action $G{:}A\times M\longrightarrow M$ (see [Furstenberg], say, for details). For $a\in A$ the Liapunov exponents of the diffeomorphism $f(x)=A(a,x)$ are determined by the *root space decomposition* of the group. Rather than give a definition (which we would not require again), we shall consider a simple example.

Consider the maximal abelian subgroup

$$A=\left\{\begin{bmatrix} \lambda & 0 \\ 0 & \lambda^{-1}\end{bmatrix} \Big| \lambda\neq 0\right\} \subseteq G = SL(2,\mathbb{R}).$$

Let $\Gamma \subseteq G$ be a discrete subgroup such that the three dimensional manifold $M=\Gamma\backslash G/\{\pm I\}$ is compact. M is thus a compact locally symmetric space. If

$$a=\begin{bmatrix} \lambda & 0 \\ 0 & \lambda^{-1}\end{bmatrix},\ \lambda\in\mathbb{R},$$

then the associated diffeomorphism $f(\Gamma\backslash g/\{\pm I\})=f(\Gamma\backslash ag/\{\pm I\})$, where $\Gamma\backslash g/\{\pm I\}\in M$ corresponds the time-$(\log\lambda)$ geodesic flow on M and has Liapunov exponents 0, $\pm\log\lambda$.

Notes

Measure theoretic entropy was introduced by Kolmogorov in 1958 (and the definition was refined by Sinai in the following year) as an isomorphism invariant. The definition of measure theoretic entropy we have adopted is taken from Katok's 1980 article [Katok].

The concept of topological entropy was introduced in 1965 by Adler, Konheim and McAndrew, with equivalent definitions being given later by Bowen and Dinaburg.

The variational principal which relates the two types of entropy was established by Goodwyn in 1970 (extending an earlier result of Parry in certain special cases).

The Pesin-Ruelle inequality has its origins in Pesin's 1976 proof of an equality for C^2 volume preserving definitions (which in turn developed previous work of Margulis and S. Katok). Subsequently, Ruelle observed that the proof gave an *inequality* for any invariant measure (this observation was published in 1978 in [Ruelle_2]).

Ergodic theory has probably found one of its most significant applications in Lie theory. A comprehensive account is contained in the books of Margulis and Zimmer referred to earlier.

Chapter 4

The Pesin set and its structure

In the previous two chapters we have introduced the Liapunov exponents of an ergodic measure $\mu \in \mathcal{M}_{erg}$ which describe the asymptotic behavior of the iterates of a C^1 diffeomorphism $f: M \longrightarrow M$ and estimated these numbers in terms of the entropies $h_{meas}(\mu)$ and $h_{top}(f)$. In this chapter we want to go further into understanding the 'structure' of the actual set of points in M for which the asymptotic behavior of f^n is defined by these Liapunov exponents.

Definition. We call an ergodic measure $m \in \mathcal{M}_{erg}$ a *hyperbolic measure* if:

(i) none of the Liapunov exponents for m are zero; and

(ii) there exist Liapunov exponents with different signs.

We denote the subset of ergodic hyperbolic measures by $\mathcal{M}^*_{erg} \subseteq \mathcal{M}_{erg}$.

Henceforth, we shall always make the following assumption:

Standing hypothesis I: We consider only hyperbolic measures $m \in \mathcal{M}^*_{erg}$

Important remark for surfaces. In view of the results in the previous chapter we should remember that in the particular case of interest of a surface diffeomorphism *any* ergodic measure m satisfying $h_{meas}(m) > 0$ is automatically hyperbolic.

This standing hypothesis has the advantage for us that it eliminates both the very trivial cases (e.g. atomic measures supported on periodic attractors or repellers) and some very difficult cases (e.g. examples with a zero Liapunov exponent).

4.1. Pesin set.

We want to introduce a space which has full measure with respect to our hyperbolic ergodic measure $m \in \mathcal{M}^*_{\text{erg}}$ and where the Liapunov exponents control the behavior along the orbits of points. We begin with a definition which is independent of measures:

Definition. Given $\lambda, \mu \gg \epsilon > 0$, and for all $k \in \mathbb{Z}^+$, we define $\Lambda_k = \Lambda_k(\lambda,\mu;\epsilon)$, $k \geq 1$, to be all points $x \in M$ for which there is a splitting $T_x M = E^s_x \oplus E^u_x$ with the invariance property $(D_x f^m) E^s_x = E^s_{f^m x}$ and $(D_x f^m) E^u_x = E^u_{f^m x}$ and satisfying:

(a) $\left\| Df^n | E^s_{f^m x} \right\| \leq e^{\epsilon k} e^{-(\lambda-\epsilon)n} e^{\epsilon|m|}$, $\forall m \in \mathbb{Z}$, $n \geq 1$;

(b) $\left\| Df^{-n} | E^u_{f^m x} \right\| \leq e^{\epsilon k} e^{-(\mu-\epsilon)n} e^{\epsilon|m|}$, $\forall m \in \mathbb{Z}$, $n \geq 1$; and

(c) $\tan\left(\text{Angle}(E^s_{f^m x}, E^u_{f^m x})\right) \geq e^{-\epsilon k} e^{-\epsilon|m|}$, $\forall m \in \mathbb{Z}$.

Definition. $\Lambda = \Lambda(\lambda,\mu;\epsilon) = \bigcup_{k=1}^{+\infty} \Lambda_k$ is a *Pesin set.*

We take the tangent of the angle between the bundles in (c) to allow for the 2π periodicity. Any similar trigonometric function would suffice.

Remark on interpretation. The expressions $e^{-(\lambda-\epsilon)n}$, $e^{-(\mu-\epsilon)n}$ reflect the 'strong hyperbolic' distortion around the orbit $\{f^n x\}_{n=-\infty}^{+\infty}$, of a point $x \in \Lambda$, which arise from the existence of the Liapunov exponents. The term $e^{\epsilon|m|}$ represents a small 'error' which may arise from starting at some other point $f^m x$ on the orbit of x (other than x itself). Finally, the term $e^{\epsilon k}$ is just a convenient form for an 'initial constant' associated to the hyperbolicity on each space Λ_k.

It is immediate from the definitions that these spaces are *nested*, in the sense that $\Lambda_1 \subseteq \Lambda_2 \subseteq \Lambda_3 \subseteq \cdots$. Furthermore, whilst each set Λ_k may not be f-invariant the situation may not be too bad in as much as we know that $f(\Lambda_k)$, $f^{-1}(\Lambda_k) \subseteq \Lambda_{k+1}$. Finally, we can conclude that the Pesin set Λ itself is f-invariant i.e. $f(\Lambda) = \Lambda$.

<u>*Proposition 4.1*</u>
(a) For each $k \geq 1$, Λ_k is a compact set.
(b) For each $k \geq 1$, the splitting $\Lambda_k \ni x \mapsto E^s_x \oplus E^u_x$ is continuous.

We postpone the proof to the end of this chapter.

In contrast to the situation for the sets Λ_k, $k \geq 1$, we should observe that, in general, Λ itself need *not* necessarily be compact, *nor* is it necessarily true that the splitting $E^s_x \oplus E^u_x$ is continuous on Λ.

Before proceeding further with the abstract theory, it is useful to consider a few examples.

<u>*Examples*</u>. (i) (Uniformly hyperbolic cases)
(a) Let $M = \mathbb{R}^2/\mathbb{Z}^2$ and let $f: M \longrightarrow M$ be the diffeomorphism defined by $f(x_1,x_2)+\mathbb{Z}^2 = (x_1+2x_2,\ x_1+x_2)+\mathbb{Z}^2$. At every point $x \in M$ the derivative is represented by the same matrix

$$Df = \begin{bmatrix} 2 & 1 \\ 1 & 1 \end{bmatrix} \text{ which preserves } \begin{cases} E^s_x = \text{span } \{\text{eigenvector for } \frac{3-\sqrt{5}}{2}\} \\ E^u_x = \text{span } \{\text{eigenvector for } \frac{3+\sqrt{5}}{2}\} \end{cases}$$

For this splitting of the unit tangent bundle we can explicitly compute the norm of the iterates of the tangent map using the eigenvalues to get that $\|Df^n|_{E^s}\|$, $\|Df^{-n}|_{E^u}\| = ((3-\sqrt{5})/2)^n$, where $0<(3-\sqrt{5})/2< 1$. Furthermore, we see by inspection that the angle Angle(E^s, E^u) is constant (Figure 11).

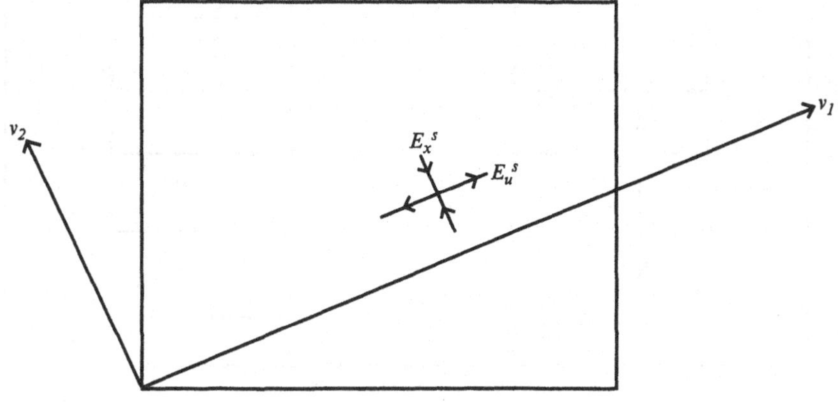

Figure 11: The Pesin set for a toral automorphism

Thus if we choose λ,μ, and ϵ such that

$$\frac{3-\sqrt{5}}{2} \leq e^{-(\lambda-\epsilon)}, e^{-(\mu-\epsilon)} < 1$$

then Λ=M is the Pesin set and we have $\Lambda_1 = \Lambda_2 = \cdots = M$.

(b) (Smale horse-shoe) We recall that the Smale horse-shoe is a diffeomorphism $f: S^2 \longrightarrow S^2$ of the 2-sphere where the interesting dynamics happen on a rectangle $R \subseteq S^2$. In particular, the recurrent part $C = \bigcap_{n=-\infty}^{+\infty} f^n R$ of R is a Cantor set. There is a natural splitting of the tangent space over each $x \in C$ given by

$$\begin{cases} E_x^s = \text{vertical direction} \\ E_x^u = \text{horizontal direction} \end{cases}$$

We recall that $0<\alpha<1/2$ denotes the contraction factor for the rectangle R. Therefore, we clearly have that $\left\|Df^n|_{E^s}\right\|, \left\|Df^{-n}|_{E^u}\right\| = \alpha^n$. Furthermore, the angle between these sections is constant (i.e. Angle$(E^s,E^u)=\pi/2$). Thus for any choice $\alpha \leq \min\{\exp-(\lambda-\epsilon), \exp-(\mu-\epsilon)\}$ we have that $C=\Lambda$ is the Pesin set with $\Lambda_k = C$, $\forall k \geq 1$ (Figure 12).

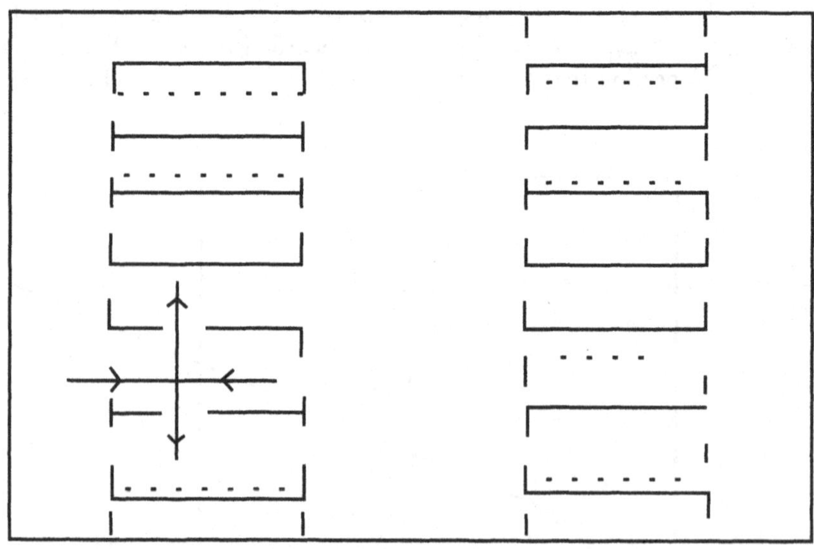

Figure 12:Pesin set for a Horse-shoe

(ii) (Non-uniformly hyperbolic cases) In section 2.4 we described how to modify the Smale horse-shoe to construct examples of non-trivial examples of Pesin sets in which are not uniformly hyperbolic. We briefly return to these examples to identify the associated Pesin sets.

(a) (A modified horse-shoe). Figure 13 shows where the expansion in the bundle E^u is uniform, but the contraction rate $\alpha(x)$ in the bundle E^s approaches unity at a single fixed point (cf. Figure 6). With the exception of this fixed point, all other points spend a bounded proportion of their orbits where E^s is uniformly contracted. This ensures that we can choose $\Lambda = C-\{\textit{fixed point}\}$, for appropriate $\lambda,\mu,\epsilon\neq 0$.

(b) (Another modified horse-shoe). Figure 14 shows where both the contraction rate $\alpha(x)<1$ and the expansion rate $\beta(x) >1$ approaches unity at a fixed point. With the exception of this fixed point, all other points spend a bounded proportion of their orbits where E^s and E^u are uniformly contracted and expanded, respectively. This ensures that we can choose $\Lambda = C-\{\textit{fixed point}\}$, for appropriate $\lambda,\mu,\epsilon\neq 0$.

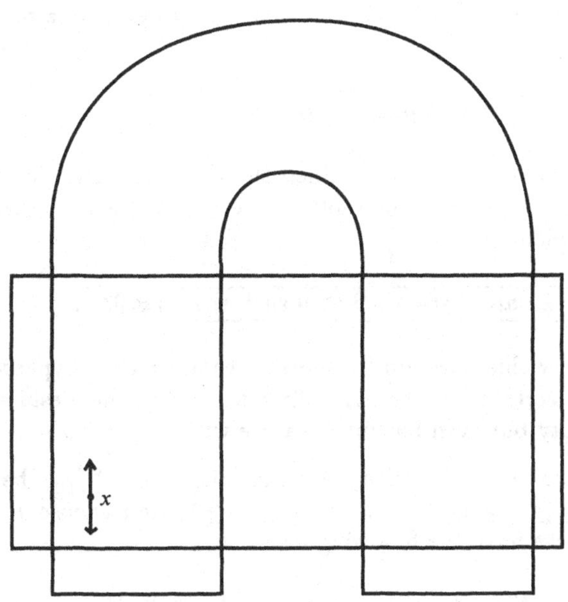

Figure 13: Contraction tends to unity at a fixed point

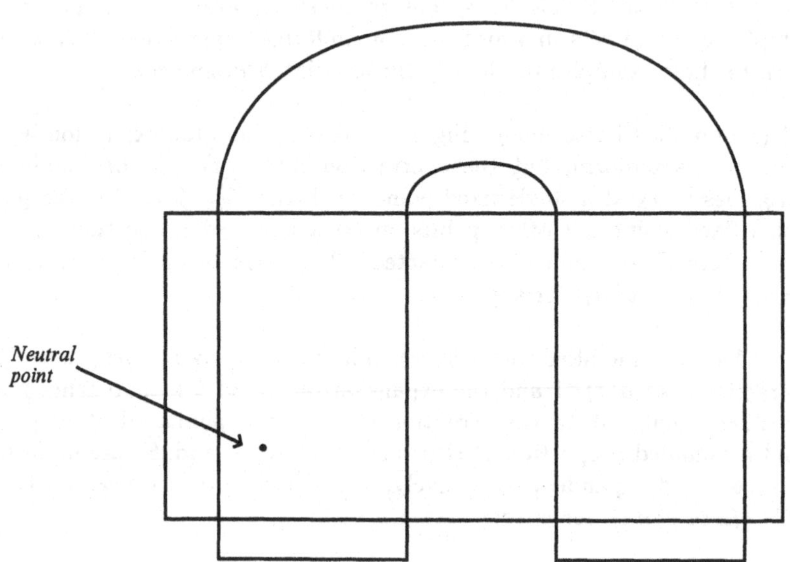

Figure 14: Contraction and expansion tends to unity at a fixed point

4.2. The Pesin set and Liapunov exponents.

Although we have confidently defined the Pesin set Λ we have yet to give a criterion for it to be non-empty. We formulate this question as the following problem

Problem. When are there $\lambda,\mu>\epsilon>0$ such that $\Lambda\neq\emptyset$?

We will answer this question by showing that, given a hyperbolic measure m, there are certain choices $\lambda,\mu>\epsilon>0$ for which the Pesin set Λ is not only non-empty but even has full measure with respect to m.

To be more precise, assume that $m\in\mathcal{M}^*_{\text{erg}}$ has Liapunov exponents $\lambda_1\geq\cdots\geq\lambda_r > 0 > \lambda_{r+1} \geq\cdots\geq\lambda_k$ and choose $\mu = \lambda_r$, $\lambda = |\lambda_{r+1}|$. Then we have the following result.

Proposition 4.2 With the above choices of λ, $\mu>0$, $\forall\epsilon$ such that λ, $\mu > \epsilon > 0$ we have $m(\Lambda) = 1$.

We postpone the proof of Proposition 4.2 to the end of the chapter. In particular, we get the following solution to the previous problem.

Corollary 4.2.1 There exist constants λ, $\mu > 0$, such that for all ϵ satisfying $\lambda,\mu>\epsilon>0$ we have that $\Lambda \neq \phi$ with these choices.

4.3 Liapunov metrics on the Pesin set.

Assuming that we have chosen a non-empty Pesin set Λ we want to change the metric on Λ so that $f\colon \Lambda \longrightarrow \Lambda$ is *uniformly* hyperbolic. In particular, we want to replace the induced metric $\|\ \|$ on $T_\Lambda M$ by a new metric $\|\ \|'$ which removes the need for constants $e^{\epsilon k}$ in measuring the 'strong hyperbolicity' on each set Λ_k. Clearly this must involve some loss of accuracy. The cost is that the two metrics $\|\ \|$ and $\|\ \|'$ become more difficult to compare on Λ_k as k increases. Let us now be specific. Choose $\epsilon>0$ sufficiently small that

$$\begin{cases} \lambda'=\lambda-2\epsilon>0 \\ \mu'=\mu-2\epsilon>0 \end{cases}$$

We can begin by defining new norms $\|\ \|_s$, $\|\ \|_u$ on the spaces E^s_x, E^u_x, respectively.

$$\begin{cases} \text{For } v_s \in E^s_x \text{ we define } \|v_s\|_s = \sum_{n=0}^{+\infty} e^{\lambda' n}\|D_x f^n(v_s)\| \\ \text{For } v_u \in E^u_x \text{ we define } \|v_u\|_u = \sum_{n=0}^{+\infty} e^{\mu' n}\|D_x f^{-n}(v_u)\| \end{cases} \tag{4.1}$$

Observe that the two series in (4.1) converge. For example, if $x\in\Lambda_k$ then

$$\begin{aligned} \sum_{n=0}^{+\infty} e^{\lambda' n}\|D_x f^n(v_s)\|_s &\leq \sum_{n=0}^{+\infty}\left(e^{\lambda' n}e^{-(\lambda-\epsilon)n}\right)e^{\epsilon k}\|v_s\|_s \\ &\leq \left(\sum_{n=0}^{+\infty} e^{-\epsilon n}\right)e^{\epsilon k}\|v_s\|_s < +\infty \end{aligned} \tag{4.2}$$

(and similarly for the other series).

We next define a new norm $\|\ \|'$ on $T_\Lambda M$ by $\|v\|'_x = \max\left\{\|v_s\|_s, \|v_u\|_u\right\}$, where $v=v_u+v_s \in E^u_x\oplus E^s_x$. This metric $\|\ \|'_x$ exhibits a 'uniform hyperbolicity' which is independent of $x\in\Lambda_k$. For example,

$$\|D_x f(v_s)\|' = \sum_{n=1}^{+\infty} e^{\lambda' n} \|D_x f^{n+1}(v_s)\|$$
$$\leq e^{-\lambda'} \Big(\sum_{n=0}^{+\infty} e^{\lambda' n} \|D_x f^n(v_s)\| \Big)$$
$$= e^{-\lambda'} \|v_s\|', \; \forall \; v_s \in E_x^s \qquad (4.3)_1$$

Similarly, we can show

$$\|D_x f^{-1}(v_u)\|' \leq e^{-\mu'} \|v_u\|', \; \forall \; v_s \in E_x^u \qquad (4.3)_2$$

Although the new metric $\| \; \|'$ is clearly well suited for f, the 'price paid' is that the two metrics $\| \; \|$, $\| \; \|'$ can differ quite significantly as k increases. Specifically, we have the worsening estimate

$$\frac{1}{\sqrt{d}} \|v\|_x \leq \|v\|_x' \leq C \, e^{\epsilon k} \|v\|_x, \; \forall x \in \Lambda_k \qquad (4.4)$$

for some constant $C>0$ independent of k (and where $d=\dim M$ denotes the dimension of M) . The estimate from above in (4.4) comes directly from (4.2) with $C = 2\sum_{n=0}^{+\infty} e^{-\epsilon n} = 2/(1-e^{-\epsilon})$, which is independent of $k \geq 1$. The lower bound comes from some easy linear algebra; see [Fa-He-Yo].

We can summarize these estimates as follows.

Proposition 4.3 There exists a metric $\| \; \|_x'$ on $T_x M$, $x \in \Lambda$, such that

(i) $\|D_x f|_{E_x^s}\|' \leq e^{-\lambda'}$, $\|D_x f^{-1}|_{E_x^u}\|' \leq e^{-\mu'}$, and

(ii) $\frac{1}{\sqrt{d}} \|v\|_x \leq \|v\|_x' \leq C \, e^{\epsilon k} \|v\|_x$, $\forall v \in T_x M$, $x \in \Lambda_k$.

Definition. We call $\| \; \|_x'$ a *Liapunov metric.*

Remark. The above construction is very similar in spirit to a part of the work of Mather on Axiom A diffeomorphisms (which would correspond to the case $\Lambda = \Lambda_k$, for some $k \geq 1$; see Interlude) in which he showed the existence of 'adapted metrics' (i.e. metrics for which the constant term can be taken to be unity; cf. [Mather].)

4.4 Local distortion.

If $x \in \Lambda$ then we want to extend the hyperbolic behavior of $D_x f^n$ to a small neighborhood of the orbit of x. In practice, this requires making a slightly stronger regularity assumption on $f: M \longrightarrow M$. So far we have only assumed that f was C^1. This means that the derivative $x \mapsto D_x f \in L(T_x M, T_{fx} M)$ is continuous (where by taking local charts we can represent $D_x f$ as a matrix in $L(\mathbb{R}^d, \mathbb{R}^d)$, where $d = \dim M$; see Appendix B). Given $\alpha > 0$ we say that $f: M \longrightarrow M$ is $C^{1+\alpha}$ if f is C^1 and furthermore the derivative $x \mapsto D_x f$ is *α-Hölder continuous* i.e. $\exists K > 0$ such that $\|D_y f - D_x f\| \leq K[d(y,x)]^\alpha$, whenever y is sufficiently close to x.

We now want to take this as a basic hypothesis.

<u>*Standing hypothesis*</u> <u>II</u>: f is of class $C^{1+\alpha}$, for some $\alpha > 0$.

<u>*Remark.*</u> It is easy to see that since M is compact a sufficient condition for this hypothesis is that $f: M \longrightarrow M$ is C^2.

We begin by fixing a point $x \in \Lambda$. By taking charts about $x, f(x) \in M$ we can assume without loss of generality that $x \in \mathbb{R}^d$, $fx \in \mathbb{R}^d$. We can simplify our analysis as follows.

(a) For a sufficiently small neighborhood $x \in \mathcal{U} \subset M$ we can trivialize the tangent bundle over $\mathcal{U}$ by identifying $T_{\mathcal{U}} M \equiv \mathcal{U} \times \mathbb{R}^d$. For convenience we can write $d(x,y) = |x - y|$.

Consider the Liapunov norm $\| \; \|'$ on the fibers $T_x M$ (over x) and $T_{fx} M$ (over fx). For any point $y \in \mathcal{U}$ and tangent vector $v \in T_y M$ we can then use the identification $T_{\mathcal{U}} M \equiv \mathcal{U} \times \mathbb{R}^d$ to 'translate' the vector v to a corresponding vector $\overline{v} \in T_x M$. We then define $\|v\|_y'' = \|\overline{v}\|_x'$. This defines a new norm $\| \; \|''$ on $T_{\mathcal{U}} M$ (which agrees with $\| \; \|'$ on the fiber $T_x M$). Similarly, we can define $\| \; \|_z''$ on $T_z M$ (for any z in a sufficiently small neighborhood $\mathcal{V}$ of fx).

We can also define a new splitting $T_y M = E_y^{s\prime} \oplus E_y^{u\prime}$ (where $y \in \mathcal{U}$) by 'translating' the splitting $T_x M = E_x^s \oplus E_x^u$ (and similarly for $T_z M = E_z^{s\prime} \oplus E_z^{u\prime}$).

(b) We want to quantify the 'distortion' of Df with respect to this new norm $\| \; \|''$. Given a tangent vector $(y,v) \in \mathcal{U} \times \mathbb{R}^d \equiv T_{\mathcal{U}} M$ with $v \in E_y^{s\prime}$, we can assume that this is the translate of a tangent vector $(x,\overline{v})\in\{0\}\times\mathbf{R}^d\equiv T_xM$ with $\overline{v}\in E^s_x$. Then,

$$\|D_yf(v)\|''_{f(y)} := \|D_yf(\overline{v})\|'_{f(x)} = \|D_xf(\overline{v})+D_yf(\overline{v})-D_xf(\overline{v})\|'_{f(x)}$$
$$\leq \|D_xf(\overline{v})\|'_{f(x)} + \|D_yf(\overline{v}) - D_xf(\overline{v})\|'_{f(x)} \tag{4.5}$$

Assume that $x\in\Lambda_k$ (and, therefore, $fx\in\Lambda_{k+1}$). Since $\overline{v}\in E^s_x$ we can apply (4.3) to the first term of (4.5) and apply (4.4) (with the $C^{1+\alpha}$ hypothesis) to the second term of (4.5) to get

$$\begin{aligned}\|D_yf(v)\|''_{f(y)} &\leq e^{-\lambda'}\|\overline{v}\|'_x + C\,e^{\epsilon(k+1)}\|D_yf(\overline{v})-D_xf(\overline{v})\|_{f(x)}\\ &\leq e^{-\lambda'}\|\overline{v}\|'_x + C\,e^{\epsilon(k+1)}\left(K\,|y-x|^{\alpha}\right)\|\overline{v}\|_x\\ &\leq \left(e^{-\lambda'} + C\,e^{\epsilon(k+1)}\,K\,|y-x|^{\alpha}\sqrt{d}\right)\|\overline{v}\|'_x \end{aligned} \tag{4.6}$$

(where in the last line we have used the other inequality from (4.4)).

(c) Choose any $0<\lambda''<\lambda'$ and then define

$$\epsilon^s_k = \min\left\{1, \left(\frac{e^{-\lambda''} - e^{-\lambda'}}{C\,e^{\epsilon(k+1)}K\sqrt{d}}\right)^{\frac{1}{\alpha}}\right\} > 0$$

We can conclude from (4.6) that whenever $|y-x|<\epsilon^s_k$ we have that

$$\|D_yf(v_s)\|''_{f(y)} \leq e^{-\lambda''}\,\|v_s\|''_y, \text{ for any } v_s \in E^{s\prime}_y.$$

A similar series of estimates allows us to deduce that, for any choice $0<\mu''<\mu'$, we can choose $\epsilon^u_k>0$ such that whenever $|y-x|<\epsilon^u_k$ then

$$\|D_yf^{-1}(v_u)\|''_{f^{-1}(y)} \leq e^{-\mu''}\,\|v_u\|''_y, \text{ for any } v_u\in E^{u\prime}_y.$$

We can now conveniently summarize these estimates as follows.

Conclusion. There exist $0<\lambda''<\lambda$, $0<\mu''<\mu$ and $\epsilon_0>0$ such that if we set $\epsilon_k=e^{-\epsilon k}\epsilon_0$ then for any point $y\in B(x,\epsilon_k)$ in an ϵ_k neighborhood of $x\in\Lambda_k$ we have a splitting $T_yM = E^{s\prime}_y\oplus E^{u\prime}_y$ with the following 'hyperbolic behavior' for D_yf

$$\begin{cases} \|D_y f(v_s)\|''_{f(y)} \le e^{-\lambda''} \|v_s\|''_y \\ \|D_y f^{-1}(v_u)\|''_{f^{-1}(y)} \le e^{-\mu''} \|v_u\|''_y \end{cases} \tag{4.7}$$

(The constant ϵ_0 depends on various 'global' properties of f, e.g. the Holder constants, the size of the local trivializations, etc.)

4.5 Proofs of Propositions 4.1 and 4.2.

We return to the previously omitted proofs of Proposition 4.1 and Proposition 4.2.

Proof of Proposition 4.1 (This appears as Theorem (1.3.1) on p.1271 of [Pesin$_2$].) To prove part (a) we need to show that Λ_k is closed.

Step 1. The hyperbolicity implies that the splitting $T_xM = E^s_x \oplus E^u_x$ is unique (because of the hyperbolic nature of the splitting, exactly as in the Axiom A case).

Step 2. If $x \in M$ and $x_i \in \Lambda_k$ $(i \ge 1)$ is a convergent sequence with $x_i \longrightarrow x$, as $i \longrightarrow +\infty$, with the choice of k fixed, then by a compactness argument we can choose a convergent subsequence of the subspaces $E^u_{x_{i_j}} \longrightarrow E^u_x$, $E^s_{x_{i_j}} \longrightarrow E^s_x$. This defines E^u_x and E^s_x.

Step 3. In the original definition of the Pesin sets we can first take $m=0$. Since by assumption $x_{i_j} \in \Lambda_k$, conditions (a), (b), (c) are satisfied by $E^u_{x_{i_j}}$, $E^s_{x_{i_j}}$, $\forall n \ge 1$. If we now let $j \longrightarrow +\infty$ we get that (a), (b), (c) are satisfied for E^u_x, E^s_x by continuity.

Step 4. It remains to deal with the cases $m \ne 0$. We want to define $E^u_{f^m x}$ to be $D_x f^m(E^u_x)$. Fix $m \in \mathbb{Z}$; then by continuity we know that as $x_{i_j} \longrightarrow x$, $D_{x_{i_j}} f^m \longrightarrow D_x f^m$. In particular, this gives

$$E^u_{f^m x_{i_j}} = D_{x_{i_j}} f^m (E^u_{x_{i_j}}) \longrightarrow D_x f^m (E^u_x) := E^u_{f^m x}.$$

The conditions (a),(b), and (c) are satisfied for $E^u_{f^m x_{i_j}}$ (for each fixed $m \in \mathbb{Z}$).

Letting $j \longrightarrow +\infty$ we get that (a), (b), (c) are satisfied for $E^u_{f^m x}$

and $E^s_{f^m x}$ by continuity. In particular, we can conclude that $x \in \Lambda_k$ and therefore that Λ_k is *closed*.

To prove part (b) we have to know that the splitting $\Lambda_k \ni x \mapsto E^s_x \oplus E^u_x$ is *continuous*. However, this is a direct consequence of the uniqueness condition (Step 1), that there is only one possible limit for $E^u_{x_{i_j}}$ and $E^s_{x_{i_j}}$. □

Proof of Proposition 4.2 (see Theorem (1.2.1) on p.126 of [Pesin$_1$]). Let m be an ergodic hyperbolic measure. For notational simplicity we shall only consider the case of surfaces i.e. $\dim M = 2$, and then we can also assume that there are Liapunov exponents $\lambda_1 > 0 > \lambda_2$.

Step 1. By the Oseledec theorem (Theorem 2.1), we know that there exists a measurable subset $\Omega \subseteq M$ with full measure i.e. $m(\Omega) = 1$, such that, whenever $x \in \Omega$,

$$\forall v \in E^i_x \ (v \neq 0), \qquad \lim_{n \to \infty} \frac{1}{n} \log \left\| D_x f^n|_{E^i_x} v \right\| \longrightarrow \lambda_i \qquad \text{(for } i=1,2)$$

If we fix $\epsilon > 0$ then for each $x \in \Omega$ there exists $N(x) > 0$ such that

$$\forall n \geq N(x), \qquad e^{(\lambda_i - \epsilon)n} \leq \left\| D_x f^n|_{E^i_x} \right\| \leq e^{n(\lambda_i + \epsilon)}$$

We begin with $i=2$. Given $x \in \Omega$ we define $C(x) > 0$ to be the *smallest* constant satisfying

$$\left\| D_x f^n|_{E^2_x} \right\| \leq C(x)\, e^{(\lambda_2 + \epsilon)n}, \quad \left\| D_x f^n|_{E^2_x} \right\| \geq \frac{1}{C(x)} e^{(\lambda_2 - \epsilon)n}, \quad \forall n \geq N(x) \tag{4.8}$$

Step 2. Choose integers $m, n > 0$ then by the chain rule we have $D_{f^m x} f^n = (D_x f^{n+m}).(D_{f^m x} f^{-m})$ and $(D_{f^m x} f^{-m}).(D_x f^m) = I$. Therefore,

$$\left\| D_{f^m x} f^n|_{E^2_{f^m x}} \right\| \leq \left\| D_x f^{n+m}|_{E^2_x} \right\| \cdot \left\| D_{f^m x} f^{-m}|_{E^2_{f^m x}} \right\|$$

$$\leq \frac{\left\| D_x f^{n+m}|_{E^2_x} \right\|}{\left\| D_x f^m|_{E^2_x} \right\|} \tag{4.9}$$

Substituting (4.8) into (4.9) gives

$$\left\| D_{f^m x} f^n |_{E^2_{f^m x}} \right\| \leq C(x)^2 \frac{e^{(\lambda_2+\epsilon)(n+m)}}{e^{(\lambda_2-\epsilon)m}} = C(x)^2 \, e^{(\lambda_2+\epsilon)n} \, e^{2\epsilon m}$$

Step 3. Since $C(x)$ was chosen to be the smallest value satisfying (4.8) we can deduce from the last inequality that $C(f^m x) \leq e^{2\epsilon|m|} C(x)^2$ (where there is a similar argument to deal with the case $m<0$).

If we choose $k_2 \geq 1$ so that $e^{\epsilon k_2} \geq C(x)^2$ (and also bounds the terms for $n=0,\ldots,N(x)-1$) then $x \in \Omega$ satisfies condition (a) in the definition of Λ_{k_2} (where we should take $E^2_x = E^s_x$, $\lambda_2 = -\lambda$ and we must replace ϵ by $\epsilon/2$, etc.).

Similar estimates replacing E^2_x with E^1_x (and k_2 replacing k_1) give that $x \in \Omega$ satisfies condition (b) in the definition of Λ_{k_2} (where this time around we should take $E^1_x = E^u_x$, $\lambda_1 = -\lambda$ and we must replace ϵ by $\epsilon/2$, etc.). We need only choose $k \geq \max\{k_1, k_2\}$ to get (a) and (b) simultaneously satisfied.

Step 4. All that remains is to show that condition (c) holds. Let $\gamma(x) = \tan\left(\mathrm{Angle}(E^s_x, E^u_x)\right)$. It follows from the detailed proof of the Oseledec theorem that $\lim_{n \to +\infty} \frac{1}{n} \log \gamma(f^n x) = 0$, $x \in \Omega$ (cf. [Pesin$_{1,2}$]). In particular, given $\epsilon > 0$ there exists an integer $N(x) \geq 1$ such that

$$e^{-\epsilon n} \leq \gamma(f^n x), \quad \forall \, n \geq N(x)$$

(i.e we choose $N(x)$ sufficiently large that $\forall n \geq N(x)$, $|\log \gamma(f^n x)| \leq \epsilon n$). We choose $B(x) \geq 1$ to be the *largest* constant such that

$$\gamma(f^n x) \geq B(x) \, e^{-\epsilon n}, \quad \forall n \geq N(x) \tag{4.10}$$

We can make two observations.

(1) If we substitute $f^m x$ for x in (4.10) we get:

$$\gamma(f^{n+m} x) \geq B(f^m x) \; e^{-\epsilon n} \tag{4.11}$$

(2) If we substitute $n+m$ for n in (4.10) we get:

$$\gamma(f^{n+m} x) \geq B(x) \; e^{-\epsilon(n+m)} \tag{4.12}$$

Finally, we can compare (4.11) and (4.12) and recall that $B(x)$ was chosen as the largest value satisfying (4.10). Thus we require that $B(f^m x) \geq B(x)\, e^{\epsilon m}$ (and similarly for $m<0$). Therefore, if we choose k sufficiently large that $e^{-\epsilon k} \leq B(x)$ (and also to deal with the cases $n=0,\ldots,N(x)$) then we see that condition (c) of Section 4.1 holds.

We conclude that $x \in \Lambda_k$ (with k chosen as above) and this completes the proof of Proposition 4.2. □

4.6 Liapunov exponents with the same sign.

We have denoted by $\mathcal{M}^*_{\text{erg}}$ those ergodic measures whose Liapunov exponents are non-zero *and* of different signs. The assumption that all of the Liapunov exponents are non-zero is important to us since otherwise we are led into complicated situations without hyperbolic behavior. By contrast, measures which fail on the second point tend to be very easy to deal with, so we shall quickly deal with them now.

Proposition 4.4 If all Liapunov exponents of $m \in \mathcal{M}_{\text{erg}}$ are non-zero and have the same sign then m is concentrated on a single orbit.

Proof. Assume without loss of generality that all the Liapunov exponents of $f: M \longrightarrow M$ are strictly negative (otherwise we can replace f by its inverse f^{-1}; see Remark (ii), Section 2.2).

For any point $x \in \Lambda$ which is a density point for the ergodic measure m we can show x is actually *periodic* (where we call x a *density point* for the measure m if $\forall \eta > 0$, $m(D_\epsilon(x,d)) > 0$). Assume that $x \in \Lambda_k$ and choose a sufficiently small (Liapunov) neighborhood $B(x,\epsilon_k)$. Since x is a density point, we have $m(B(x,\epsilon_k)) > 0$ and by Poincaré recurrence $f^n B(x,\epsilon_k) \cap B(x,\epsilon_k) \neq \emptyset$, for some $n>0$. Since all the Liapunov exponents are negative, the map $f^n: B(x,\epsilon_k) \longrightarrow f^n B(x,\epsilon_k)$ is contracting and, with the previous observation, we can suppose that $f^n: B(x,\epsilon) \longrightarrow B(x,\epsilon)$ is a contraction, for some appropriate $\epsilon>0$.

By the contraction mapping theorem $\bigcap_{n=0}^{+\infty} f^n B(x,\epsilon)$ contains exactly one fixed point for f^n. However, since the measure m is *preserved* by f^n, and every neighborhood of m has non-zero measure, we conclude that $f^n x = x$ is the unique point in this intersection and $m(B(x,\epsilon)) = m(\{x\})$.

So far we have only used the *invariance* of the measure m to deduce that each density point $x \in M$ is periodic and has non-zero m-measure. However, *ergodicity* of $m \in \mathcal{M}_{\text{erg}}$ implies that m is supported on a single orbit. □

Notes

The Pesin set was introduced in Pesin's 1976 articles [$\text{Pesin}_{1,2}$]. Our approach takes a slightly more topological viewpoint, in common with the approach of Katok and Newhouse. The idea of a Liapunov metric is also introduced in Pesin's article, although we follow more closely the survey article of Fathi-Herman-Yoccoz **[Fa-He-Yo]**. The estimates on local distortion that we give are adapted from Newhouse's 1989 article [$\textbf{Newhouse}_1$], where the presentation is a little more concrete than in other works. However, the proofs in Section 4.5 are more or less taken from Pesin's original articles.

An interlude

In the first four chapters (constituting Part I of these notes) we have developed the technical machinery we shall need for the applications in Part II. However, before continuing we shall take the opportunity to discuss a couple of topics which, although not strictly necessary for our account, do help to shed a little light on matters.

(a) Some topical examples.

We want to mention briefly a couple of examples to which the general ideas of Pesin theory can be applied. Both examples deal with diffeomorphisms of low-dimensional manifolds, and have enjoyed a certain amount of recent interest. We give references where more information can be found.

Example 1 (The Hénon map). In recent years the Hénon map has been the subject of study as one of the principal examples of so called 'strange attractors'. For our purposes, we shall define a ***strange attractor*** associated to a diffeomorphism $f: M \longrightarrow M$ to be a compact f-invariant subset $\Lambda \subset M$ satisfying the following properties:

(i) $f: \Lambda \longrightarrow \Lambda$ is transitive (i.e there exists a dense orbit);

(ii) the basin of attraction $W^s(\Lambda) = \{x \in M \mid d(f^n x, \Lambda) \longrightarrow 0 \text{ as } n \longrightarrow +\infty\}$

has non-empty interior;

(iii) Λ supports a probability measure with non-zero Liapunov exponents.

Remark. Usually, we do not include the simple cases of uniformly hyperbolic attractors or attracting periodic orbits within this definition.

The Hénon maps are a family of polynomial maps $f_{a,b}: \mathbb{R}^2 \longrightarrow \mathbb{R}^2$ of the plane given by

$$f_{a,b}\begin{bmatrix} x \\ y \end{bmatrix} = \begin{bmatrix} y+1-ax^2 \\ bx \end{bmatrix}$$

(where $a,b \in \mathbf{R}$ are two real parameters). Hénon performed experiments using computer simulations of the iterates of the maps with the choice of values $a=1.4$ and $b=0.3$, which strongly suggested that for these parameters there exists a strange attractor (cf. Figure 15, and see [Ruelle$_3$], for a particularly nice account).

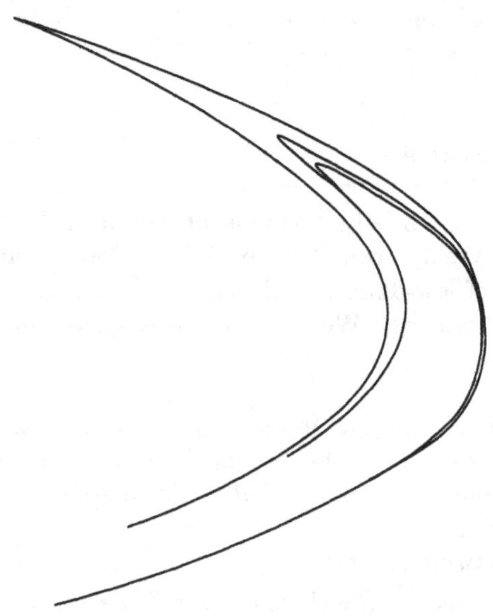

Figure 15: Iterates of (0,0) under the map $f_{a,b}$ with $a=1.4$, $b=0.3$

Unfortunately, despite the simplicity of these equations, rigorous proofs of the existence of strange attractors are very difficult. As the culmination of an immense program, Benedicks and Carleson have shown the following. *For $b>0$ sufficiently small there is a set of $a>0$ values of positive Lebesgue measure for which $f_{a,b}$ has a strange attractor* [Ben-Car]. The reason that $b>0$ should be small is that the starting point for this analysis is the consideration of perturbations of the map $f_{a,0}$ (see Figure 16) whose dynamics reduce to those of a one-dimensional quadratic map, whose dynamics, while not particularly simple, have been studied by Jacobson, Benedicks-Carleson, and others; see [Whitley] for a brief survey of this approach.

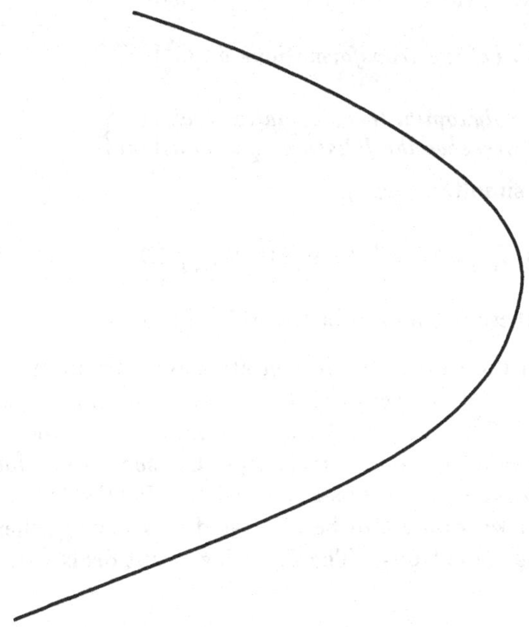

Figure 16: Iterates of (0,0) under $f_{2,0}$ and $f_{2,0.2}$

Although the ambient manifold $\mathbb{R}^2$ is not compact, the interesting dynamics take place on the compact set Λ. These diffeomorphisms have non-zero topological entropy since it is easy to show that there exist homoclinic tangencies with in Λ, and thus (generalized) Smale horse-shoes of strictly positive topological entropy. In particular, we can apply most of the theory described in this text.

The Hénon maps are examples of polynomial maps on $\mathbb{R}^2$, although they obviously define polynomial maps on $\mathbb{C}^2$ (just by considering $(x,y) \in \mathbb{C}^2$). Our second class of examples generalize these extensions.

Example 2 (Polynomial diffeomorphisms on $\mathbb{C}^2$). We can consider quite general polynomial diffeomorphisms $f: \mathbb{C}^2 \longrightarrow \mathbb{C}^2$ of the form $f(z_1,z_2) =(P(z_1,z_2), Q(z_1,z_2))$, where $P,Q \in \mathbb{C}[z_1,z_2]$ are polynomials in the two variables $z_1,z_2 \in \mathbb{C}$. We define the degree of f to be the maximum of the degrees of the two components (i.e. $d = \deg(f) := \max\{\deg(P), \deg(Q)\}$).

It has been shown that every polynomial f diffeomorphism of $\mathbb{C}^2$ can be factorized in the form $f=f_1\circ\cdots\circ f_k$, where

$$\text{(i)}\begin{cases}\text{either } f_i\in A=\{\textit{affine transformations on } \mathbb{C}^2\} \\ \text{or } f_i\in E=\left\{\begin{matrix}\textit{polynomial transformations which} \\ \textit{preserve the foliation } z_2 = \textit{constant}\end{matrix}\right\}\end{cases}$$

(but not both simultaneously)

(ii) If $f_i\in A$ then $f_{i+1}\in E$, and if $f_i\in E$ then $f_{i+1}\in A$

An account of these results can be found in [Fri-Mil].

As with the case of the Hénon attractor, our ambient manifold $\mathbb{C}^2$ is not compact, and so we want to restrict attention to some compact invariant set $\Lambda\subset\mathbb{C}^2$. A natural choice is for Λ to be the non-wandering set (i.e. the union of points $z=(z_1,z_2)\in\mathbb{C}^2$ such that for every open neighborhood U, $z\in U$, there exists $n\geq 1$ with $f^n U\cap U\neq\emptyset$). If we knew that Λ was compact, we would still be interested in knowing when $f\colon \Lambda\to\Lambda$ had non-zero topological entropy. The following result deals with both of these points.

Theorem A. If we assume that the factorization of $f\colon \mathbb{C}^2\longrightarrow\mathbb{C}^2$ is cyclically reduced then:

(a) the non-wandering set $\Lambda\subseteq\mathbb{C}^2$ is compact;

(b) if f has degree $d=\deg(f)$ then $f\colon \Lambda\to\Lambda$ has topological entropy $h_{\text{top}}(f) = \log(d)$

(see [Bed-Smi]).

For surfaces, the assumption that $h_{\text{top}}(f)>0$ was shown to imply the existence of ergodic measures $\mu\in\mathcal{M}_{\text{erg}}$ with non-zero Liapunov exponents $\lambda_1>0>\lambda_2$, say (see Theorem 3.2). Of course, in the present context the ambient manifold $\mathbb{C}^2$ has *four (real) dimensions.* However, since we are are considering (complex analytic) maps and the ambient manifold has *two complex dimensions* we would hope that a similar result holds. This is indeed the case, as we shall now see.

Assume that f has degree $d\geq 2$; then by part (b) of the theorem we see that $h_{\text{top}}(f)>0$. By the variational principle we can choose an ergodic measure $\mu\in\mathcal{M}_{\text{erg}}$ such that $h_{\text{meas}}(\mu)>0$. If we apply the Oseledec theorem (Theorem 2.2, applicable to diffeomorphisms of four (real) dimensional manifolds) to $\mu \in \mathcal{M}_{erg}$ then we are guaranteed the existence of four Liapunov exponents $\lambda_1, \lambda_2, \lambda_3, \lambda_4 \in \mathbb{R}$. However, since the map f is (complex) analytic these exponents must occur in pairs i.e. $\lambda_1 = \lambda_2 = \lambda$, and $\lambda_3 = \lambda_4 = \lambda'$, say. By the Pesin-Ruelle inequality (Theorem 3.1) we have that $h_{meas}(\mu)$ bounds from below the sum of the *positive* Liapunov exponents. Applying this to both f and f^{-1} we get that

$$2 \max\{0, \lambda\} + 2 \max\{0, \lambda'\} \geq h_{meas}(\mu) > 0$$

$$2 \max\{0, -\lambda\} + 2 \max\{0, -\lambda'\} \geq h_{meas}(\mu) > 0$$

and we can conclude that λ and λ' have different signs, i.e. either $\lambda_1 = \lambda_2 > 0 > \lambda_3 = \lambda_4$ or $\lambda_1 = \lambda_2 > 0 > \lambda_3 = \lambda_4$.

(This argument appears in [Fri-Mil], and is a simple variation of the proof of Theorem 3.2.)

(b) Uniformly hyperbolic systems.

Many of the results we shall consider are motivated by the special case of uniformly hyperbolic diffeomorphisms or 'Axiom A' diffeomorphisms (in the sense of Smale). These form an important class of diffeomorphisms in the space of diffeomorphisms on M which is *open* but in general *not* dense. Whereas a familiarity with the Axiom A theory is not a pre-requisite for appreciating Pesin theory, it certainly helps. Therefore, we shall briefly recall some of the main ideas and refer to the existing literature for more details.

Let $f: M \longrightarrow M$ be a C^1 diffeomorphism of a compact manifold then the non-wandering set is defined by

$$\Omega = \{x \in M \mid \forall \text{ neighborhoods } \mathcal{U} \ni x\ \exists n_i \rightarrow +\infty, f^{n_i}\mathcal{U} \cap \mathcal{U} \neq \emptyset\}$$

(i.e. the generalization of the definition for polynomial maps of $\mathbb{C}^2$). If follows immediately from this definition that Ω is closed and f-invariant.

<u>*Definition*</u>. The diffeomorphism $f: M \longrightarrow M$ is said to satisfy *Axiom A* if:

(i) the periodic points are dense in Ω;

(ii) $f: \Omega \longrightarrow \Omega$ is uniformly hyperbolic i.e. there exists a continuous Df-invariant splitting $T_\Omega M = E^s \oplus E^u$ of the tangent space with constants $C > 0$ and $\lambda, \mu > 0$ such that

$$\begin{cases} \|Df^n(v)\| \leq C e^{-\lambda n} \|v\|, \text{ for } v \in E^s \\ \|Df^{-n}(v)\| \leq C e^{-\mu n} \|v\|, \text{ for } v \in E^u \end{cases}$$

for all $n \geq 0$. This definition was introduced by Smale, and [Smale] still remains one of the best references.

Example (Smale horse-shoe). Let $f: S^2 \longrightarrow S^2$ be the diffeomorphism described before which preserves a Cantor set C contained in a rectangle $R \subseteq S^2$. The non-wandering set consists of the set C plus the two fixed points.

Remark. Let $f: M \longrightarrow M$ be any homeomorphism; then every f-invariant probability measure μ is supported on Ω (i.e. $\mu(\Omega)=1$). This is an easy result since if we assume for a contradiction that $\mu(M \backslash \Omega)>0$ then we can choose an open set $U \subset M - \Omega$ such that $\mu(U)>0$ and $U \cap f^n U = \emptyset$, $\forall n>0$. But this leads to a contradiction since $\{f^n(U) \mid n \in \mathbb{Z}\}$ are disjoint sets with the same non-zero measure, which is clearly impossible if $\mu(M) = 1$.

We shall now recall some of the more familiar properties of uniformly hyperbolic diffeomorphisms.

(i) *Shadowing*. We begin with some notation.

Definition. Given $\delta>0$ we call a sequence $\underline{x}=(x_n)_{n=-\infty}^{+\infty}$ with $x_n \in \Omega$ which satisfies $d(fx_n, x_{n+1})<\delta$, for all $n \in \mathbb{Z}$, a ***δ-pseudo-orbit***. Given $\epsilon>0$ we call $x \in \Omega$ an ***ϵ-tracing point*** for a δ-pseudo-orbit $\underline{x}$ if $d(f^n x, x_n)<\epsilon$ for all $n \in \mathbb{Z}$.

For Axiom A diffeomorphisms we have the following important result.

Proposition A (Shadowing lemma). $\forall \epsilon>0$ $\exists \delta>0$ such that every δ-pseudo-orbit has a (unique) ϵ-tracing point.

(ii) *Closing lemma*. A 'special case' of the shadowing lemma is the following *closing lemma*.

Proposition B (Closing lemma) $\forall \epsilon>0$ $\exists \beta>0$ such that whenever $d(f^n x, x) < \beta$ there exists a periodic point $f^n z = z$ with $d(x,z)<\epsilon$.

Proof of closing lemma (assuming shadowing lemma). To deduce the closing lemma from the shadowing lemma we should consider the infinite sequence $\underline{x}=(x_n)_{n=-\infty}^{+\infty}$ constructed by repeating the finite sequence $x, fx, \cdots, f^{n-1}x$. In particular, $\underline{x}$ is automatically a β-pseudo-orbit. If we had chosen β to be the value δ associated to ϵ in the shadowing lemma (Proposition A) then the same lemma tells us there is an associated (unique) shadowing point z, say, for the sequence which satisfies $d(x,z)<\epsilon$. Finally, to see that z is periodic we need only observe that by shifting any pseudo-orbit (x_n) places to the left the resulting (unique) shadowing point z is replaced by $f^n z$, but in the case of our (periodic) sequence we have the same pseudo-orbit, and therefore the same unique tracing point i.e. $f^n z = z$.

These estimates on closed orbits can be used to derive estimated results on their growth rates. Let $h_{\text{top}}(f)$ be the topological entropy of $f\colon M \longrightarrow M$.

Corollary. $\overline{\lim_{n\to+\infty}} \frac{1}{n} \log \# \{f^n x = x\} = h_{\text{top}}(f)$.

(iii) *Stable manifolds*. The splitting in the tangent bundle over Ω is reflected in the existence of ***stable and unstable manifolds*** in M. This is a useful device for transferring the hyperbolic behavior from the tangent space to the manifold.

Definition. Given $\delta>0$ we define a (local) ***stable manifold*** through $x\in\Omega$ by $W^s_\delta(x) = \{y \in M \mid d(f^n x, f^n y) \leq \delta\}$

Proposition C. There exists $\delta>0$ such that $W^s_\delta(x)$ is a C^1 embedded submanifold such that $T_x W^s_\delta(x) = E^s_x$.

If we replace f by f^{-1} then it similarly follows that $W^u_\delta(x) = \{y\in M \mid d(f^{-n}x, f^{-n}y)\leq\epsilon,\ n\geq 0\}$ is a C^1 embedded submanifold with $T_x W^u_\delta(x) = E^u_x$.

Notes

The Hénon attractor has been extensively studied by computer experiments for many years, the results of which pointed towards the existence of a strange attractor. Only in recent years has the theoretical work of Carlson and Benedicks established rigorously the existence of a strange attractor (and only then for somewhat restricted parameter values). Some more elementary properties are described in the book of Devaney [Devaney]. The study of polynomial diffeomorphisms of $\mathbb{C}^2$ follows naturally.

The idea of Axiom A diffeomorphisms originated in Smale's 1967 survey article [Smale]. The standard reference for the material on shadowing and closing lemmas is Bowen's 1975 book [Bowen]. The stable manifold theory evolved from work of Hirsch, Pugh, Shub and others in the late sixties and a nice summary occurs in Shub's book [Shub].

Chapter 5

Closing lemmas and periodic points

In the first four chapters we introduced the basic machinery of the theory. In this chapter we begin the applications of these ideas.

We shall continue to assume that $f: M \longrightarrow M$ is a $C^{1+\alpha}$ diffeomorphism of a compact manifold for which there exists a hyperbolic measure $m \in \mathcal{M}^*_{\text{erg}}$ which, as we observed in the previous chapter (see Corollary 4.2.1), implies that there exists a non-empty Pesin set Λ. As always, we shall be especially interested in the important case of surface diffeomorphisms with $h_{\text{top}}(f) > 0$.

The purpose of this chapter is to give our first application of this theory to the study of periodic points (i.e. points $x \in M$ such that $f^p x = x$, for some $p \geq 1$) and the following problem.

Problem. Are there always periodic points for $f: M \longrightarrow M$? (and how many?)

5.1 Liapunov neighborhoods.

Let us begin by considering a non-empty Pesin set Λ. Let λ, $\mu > 0$ be associated with the set Λ and let $\epsilon > 0$ be sufficiently small that

$$\begin{cases} \lambda'' = \lambda - 2\epsilon > 0 \\ \mu'' = \mu - 2\epsilon > 0 \end{cases},$$

where λ'', μ'' are chosen as in Section 4.4.

From the inequalities (4.7) in Chapter 4 we know that there exists $\epsilon_0 > 0$ such that

$$\left.\begin{aligned} \|D_y f(v_s)\|''_{fy} &\leq e^{-\lambda''} \|v_s\|''_y, \text{ for } v_s \in E^{s\prime}_y \\ \|D_y f^{1}(v_u)\|''_{f^{-1}y} &\leq e^{-\mu''} \|v_u\|''_y, \text{ for } v_u \in E^{u\prime}_y \end{aligned}\right\} \tag{5.1}$$

where the norm $\| \|''$and the splitting $E^{s\prime}_y \oplus E^{u\prime}_y$ associated to T_yM are translated from the norm $\| \|'$, and the splitting $E^s_x \oplus E^u_x$, for $x \in \Lambda_k$, and $y \in B(x,\epsilon_k)$ with $\epsilon_k = \epsilon_0.e^{-\epsilon k}$.

<u>*Definition*</u>. We define the *Liapunov neighborhood* $L = L(x,\alpha\epsilon_k)$ of $x \in \Lambda$ (with size $\alpha\epsilon_k$, $0<\alpha<1$) to be the neighborhood of x in M which is the projection onto M of the space $(-\alpha\epsilon_k,\alpha\epsilon_k)E^s_x \oplus (-\alpha\epsilon_k,\alpha\epsilon_k)E^u_x$.

Notice that this definition *does* depend on some arbitrary choices.

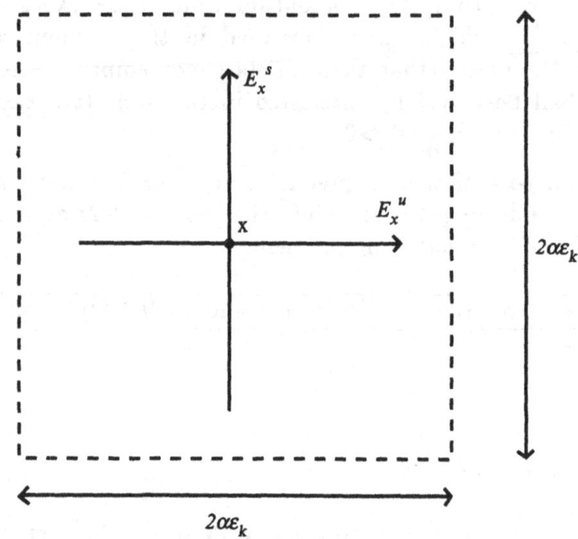

Figure 17: A Liapunov neighborhood

In particular, we see that the Liapunov neighborhood L is homeomorphic to the 'rectangle' $D^\ell \times D^n$ which is a product of discs of dimensions ℓ and n, respectively (where, $\ell = \dim E^s_x$, $n = \dim E^u_x$); see figure 17. In the particular case where the manifold M is two-dimensional we have that $\ell = n = 1$ and L is homeomorphic to a standard rectangle in the plane.

We want to use the estimates (5.1) to describe how $L(x,\alpha\epsilon_k)$ transforms under f. We begin by recalling that if $x \in \Lambda_k$ then we automatically know $fx \in \Lambda_{k+1}$ and $f^{-1}x \in \Lambda_{k+1}$ (see Section 4.1).

Lemma 5.1. The image $f(L(x,\alpha\epsilon_k))$ meets $L(fx,\alpha\epsilon_{k+1})$ transversely (in the sense that their configuration is homeomorphic to the diagram in Figure 18).

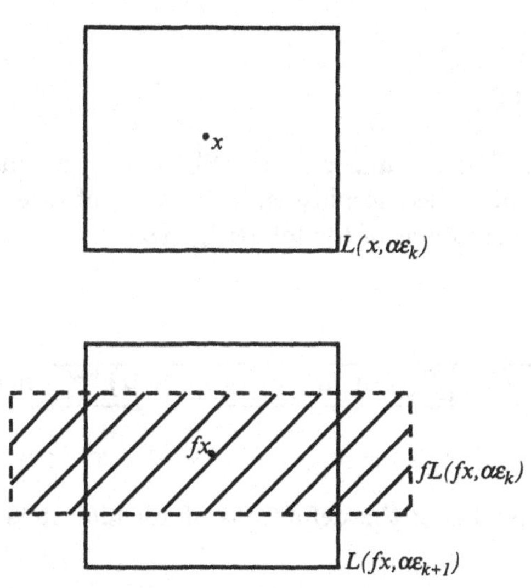

Figure 18: A transverse intersection of rectangles.

The result remains true if we replace $L(fx,\alpha\epsilon_{k+1})$ by either $L(fx,\alpha\epsilon_k)$ or $L(fx,\alpha\epsilon_{k-1})$.

Proof. There are natural local co-ordinates $(-\alpha\epsilon_k,\alpha\epsilon_k)\times(-\alpha\epsilon_k,\alpha\epsilon_k)$ on $L(x,\alpha\epsilon_k)$, and similarly on $L(fx,\alpha\epsilon_{k+1})$. With respect to these co-ordinates we have that $f(0,0)=(0,0)$. Furthermore, with these co-ordinates we can write

$$\begin{cases} L(x,\ \alpha\epsilon_k) = \{(a,b) \mid |a|,\ |b| < \alpha\epsilon_k\} \\ L(fx,\alpha\epsilon_{k+1}) = \{(c,d) \mid |c|,\ |d| < \alpha\epsilon_{k+1}\} \end{cases}$$

Using this notation we notice that if $f(a,b) = (c,d)$ then we have the inequalities $|b|\leq|d|\exp-\mu''$ and $|c|\leq|a|\exp-\lambda''$ from (5.1). The important point here is that D_xf preserves E^s_x, E^u_x and so f respects local co-ordinates. (For the second estimate we should really assume $\alpha< \exp-(\lambda''+\epsilon)$ to keep things well defined.)

Since by construction we have that $\epsilon_k/\epsilon_{k+1} = e^{\epsilon}$ it is easy to see that if we also assume, without loss of generality, that ϵ is chosen sufficiently small that $\exp(\epsilon - \lambda'')$, $\exp(\epsilon - \mu'') < 1$ (or equivalently $\lambda'', \mu'' > \epsilon$) then Lemma 5.1 easily follows from these estimates. □

5.2 Shadowing lemma.

We now want to find an analog of the Shadowing lemma described, for uniformly hyperbolic diffeomorphisms, in Part (b) of Interlude. The basic idea is to consider questions of the following type

<u>*Problem*</u>. Given $(x_n)_{n=-\infty}^{+\infty} \subset \Lambda$ find an orbit $\{f^n x\}_{n=-\infty}^{+\infty}$ nearby.

We begin by generalizing the definitions of pseudo-orbits and shadowing points.

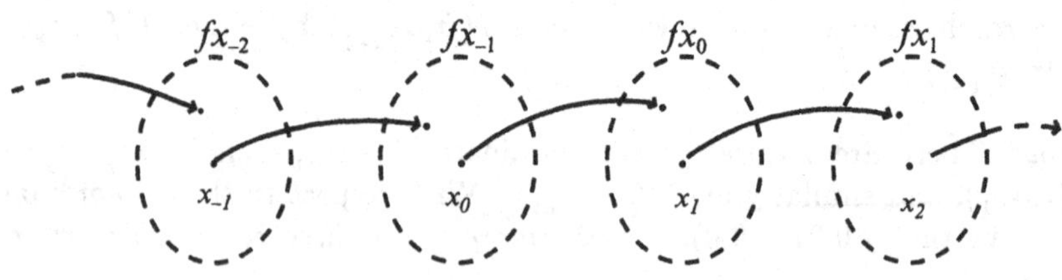

Figure 19: A pseudo-orbit

<u>*Definition*</u>. Let $(\delta_k)_{k=1}^{+\infty}$ be a sequence of positive real numbers. Let $(x_n)_{n=-\infty}^{+\infty}$ be a sequence in Λ for which there exists a sequence $(s_n)_{n=-\infty}^{+\infty}$ of positive integers satisfying:

(i) $x_n \in \Lambda_{s_n}$, $\forall\ n\in\mathbb{Z}$;

(ii) $|s_n - s_{n+1}| \leq 1$, $\forall\ n\in\mathbb{Z}$;

(iii) $d(fx_n, x_{n+1}) \leq \delta_{s_n}$, $\forall\ n\in\mathbb{Z}$

then we call $(x_n)_{n=-\infty}^{+\infty}$ a $(\delta_k)_{k=1}^{+\infty}$ - *pseudo-orbit* (see Figure 19).

Definition. Given $\eta > 0$, a point $x\in M$ is an *η-shadowing point* for the $(\delta_k)_{k=1}^{+\infty}$ pseudo-orbit if $d(f^n x, x_n) \leq \alpha\epsilon_{s_n}$, $\forall n\in\mathbb{Z}$; where $\epsilon_k = \epsilon_0 . e^{-\epsilon k}$ (see Figure 20).

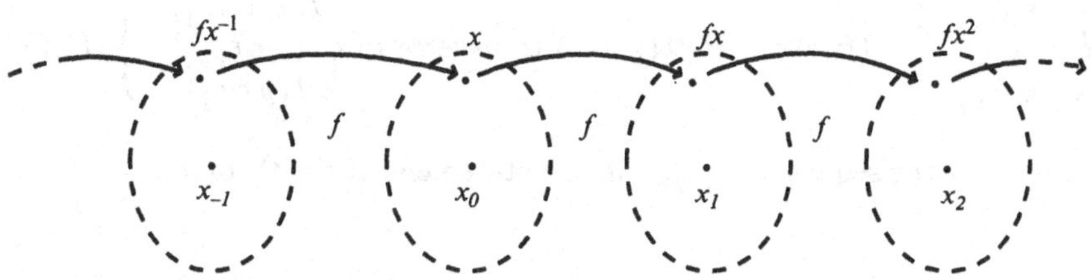

Figure 20: A shadowing point

The main result of this section is the following generalization of the shadowing lemma from the case of uniformly hyperbolic diffeomorphisms.

Theorem 5.1 (Shadowing lemma). Let $f: M\longrightarrow M$ be a $C^{1+\alpha}$ diffeomorphism with a non-empty Pesin set Λ and fixed parameters, λ, $\mu > 0$ and ϵ, $\epsilon_0 > 0$, etc. For any $\eta>0$ there exists a sequence $(\delta_k)_{k=1}^{+\infty}$ such that for *any* $(\delta_k)_{k=1}^{+\infty}$-pseudo-orbit there exists at least one η-shadowing point.

Remark. The shadowing point in Theorem 5.1 need not be an element of Λ.

Proof of Theorem 5.1. We shall first explain how the sequence $(\delta_k)_{k=1}^{+\infty}$ can be chosen and then, after a brief interruption for a necessary technical lemma, show how the shadowing point is constructed.

Step 1. (Choice of δ_k, $k\geq 1$.) For any $k\geq 1$ we know by Lemma 5.1 that the 'rectangles' $fL(x,\eta\epsilon_k)$ and $L(y,\eta\epsilon_{k+1})$ (or $L(y,\eta\epsilon_k)$ or $L(y,\eta\epsilon_{k-1})$) meet transversely when we set $y=fx$. By continuity of f and Df this transverse intersection persists if we keep x fixed but take any y sufficiently close to fx.

If we assume that $x\in\Lambda_k$ then by compactness of this set (Proposition 4.1(a)) we can chose a uniform lower bound on the size $\delta_k>0$ of neighborhoods with this 'transverse intersection' property i.e. $\forall x\in\Lambda_k$,

$$\left\{\begin{matrix} y\in\Lambda_{k-1} \\ y\in\Lambda_k \\ y\in\Lambda_{k+1}\end{matrix}\right\},\ d(fx,y)<\delta_k \Rightarrow fL(x,\alpha\epsilon_k) \text{ transverse to } \left\{\begin{matrix} L(y,\alpha\epsilon_{k-1}) \\ L(y,\alpha\epsilon_{k+1}) \\ L(y,\alpha\epsilon_k)\end{matrix}\right\} \qquad (5.2)$$

This gives the sequence $(\delta_k)_{k=1}^{+\infty}$ in the statement of the theorem.

Step 2. We next need a simple lemma which represents the analogous problem for *finite* sequences.

Lemma 5.2. Given $N>0$, let $x_n\in\Lambda_{s_n}$, $-N\leq n\leq N$, be a finite sequence satisfying $|s_n-s_{n+1}|\leq 1$ and $d(fx_n,x_{n+1})\leq\delta_{s_n}$. Then there exists $x\in M$ such that:

(a) $x\in L(x_0,\eta\epsilon_{s_0})$; and

(b)$_N$ $d(f^n x, x_n) \leq \eta\,\epsilon_{s_n}$ (i.e. $f^n x\in L(x_n,\eta\epsilon_{s_n})$, $\forall\ |n|\leq N$) (5.3)

Proof of Lemma 5.2. By construction of the sequence $(\delta_k)_{k=1}^{+\infty}$ each image 'rectangle' $fL(x_n,\eta\epsilon_{s_n})$ meets the next rectangle $L(x_{n+1},\eta\epsilon_{s_{n+1}})$ transversely (by Step 1). This is a purely 'topological' situation and so the situation is homeomorphic to the picture in Figure 21.

We can pull back the successive rectangles to a neighborhood of x and then consider the intersection defined by $I=\bigcap_{n=-N}^{N} f^{-n}L(x_n,\alpha\epsilon_{s_n})\subseteq L(x_0,\alpha\epsilon_{s_0})$. This situation will be homeomorphic to the diagram in Figure 22.

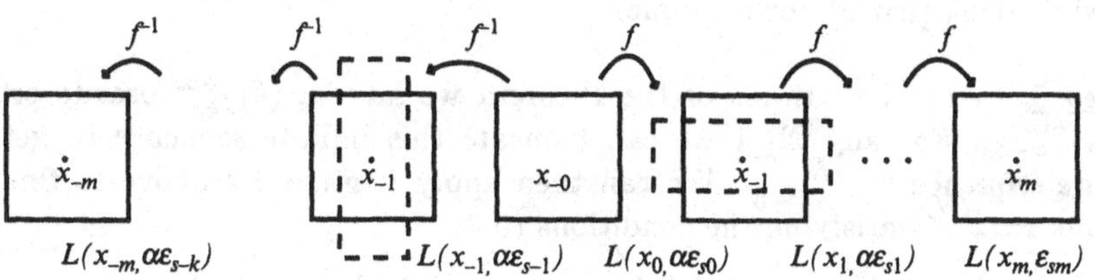

Figure 21: A sequence of transverse rectangles

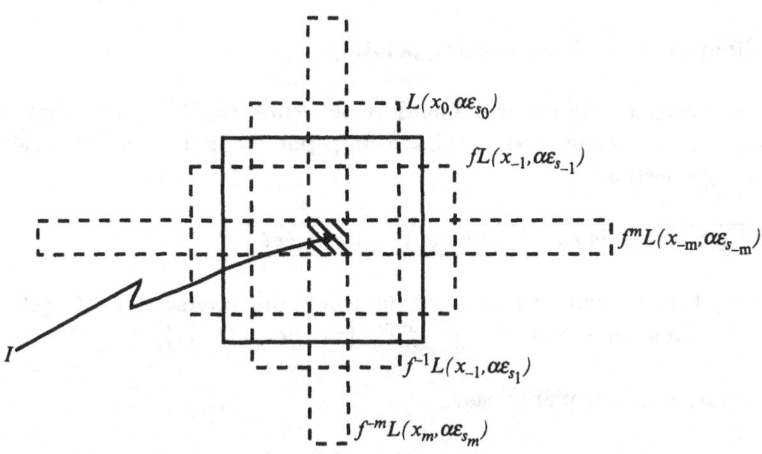

Figure 22: Intersecting rectangles

Clearly we have that this intersection is non-empty i.e. $I \neq \emptyset$. Furthermore, any point $x \in I$ automatically satisfies the required conditions. This completes the proof of the lemma. □

Note. The above type of argument is common in 'stability' results; see Irwin's book [Irwin], for example.

Step 3. In the hypothesis of the theorem we have a $(\delta_k)_{k=1}^{+\infty}$-pseudo-orbit $(x_n)_{n=-\infty}^{+\infty}$. For any $N \geq 1$ we can truncate this infinite sequence to get a *finite* sequence $(x_n)_{n=-N}^{N}$. We can then apply Lemma 5.2 above to find a point $x = x^{(N)}$ satisfying the conditions (5.3).

Since $L(x_0, \eta\epsilon_0) \subseteq M$ is closed and bounded let x be an accumulation point of $(x^{(N)})_{N=1}^{+\infty}$ in this set. In particular, from (5.3) we get that the accumulation point x satisfies:

(a) $x \in L(x_0, \eta\epsilon_{s_0})$

(b)$_\infty$ $d(f^n x, x_n) \leq \eta\epsilon_{s_n}$, $\forall\ n \in \mathbb{Z}$.

This completes the proof of the Theorem. □

5.3. Uniqueness of the shadowing point.

In the previous section we found a sequence $(\delta_k)_{k=1}^{+\infty}$ such that every $(\delta_k)_{k=1}^{+\infty}$ pseudo-orbit has a shadowing point. It remains to ask the following question:

Problem. When is the shadowing point *unique*?

Clearly, this is equivalent to asking when the intersection of rectangles consists of a single point, i.e. $\bigcap_{n=-\infty}^{+\infty} f^{-n} L(x_n, \eta\epsilon_{s_n}) = \{x\}$.

The solution is reasonably easy.

Lemma 5.3 There exists a unique shadowing point.

Proof. Using the $\|\ \|''$-norm we can deduce that the size of the set $I_N = \bigcap_{n=-N}^{N} f^n L(x_n, \eta\epsilon_{s_n})$ tends to zero as $N \longrightarrow +\infty$. In fact, we see that

$$\mathrm{diam}(I_N) \leq \epsilon_0\, (e^{-N\lambda''} + e^{-N\mu''}) \longrightarrow 0, \quad \text{as } N \longrightarrow +\infty.$$

In particular, the intersection of all the rectangles is equal to $\bigcap_{N=0}^{+\infty} I_N$ which consists of (at most) one point.

Note. Arguments of a similar nature can be found in Sinai's article [Sinai$_1$] on Markov partitions.

5.4 Closing lemmas.

As we explained in the uniformly hyperbolic case, closing lemmas are a particular type of shadowing lemmas, in which the shadowing point is a periodic point.

The main result of interest to us is the following generalized closing lemma:

Theorem 5.2 (Closing lemma, Katok). Let $f: M \longrightarrow M$ be a $C^{1+\alpha}$ diffeomorphism and let Λ be a non-empty Pesin set. For all $k \geq 1$, $0<\eta<1$, there exists $\beta=\beta(k,\eta)>0$ such that: if x, $f^p x \in \Lambda_k$ and $d(x,f^p x) < \beta$ then there exists a periodic point $f^p z = z \in M$ with $d(z,x)<\alpha$.

Proof of Theorem 5.2. Theorem 5.2 can be viewed as a corollary of Theorem 5.1. We begin by using Theorem 5.1 to give the existence of the sequence $(\delta_k)_{k=1}^{+\infty}$ specifying pseudo-orbits. The choice of β in the statement of this theorem comes from setting $\beta=\delta_{k+1}$.

Step 1. We can construct an infinite sequence $(x_n)_{n=-\infty}^{+\infty}$ from the point x as follows. We define $x_n=f^\ell x$, whenever $n=rp+\ell$, where $r\in\mathbb{Z}$, and $0\leq \ell \leq p-1$, i.e. the sequence takes the form

$$\cdots, f^{p-1}x,\ x,\ fx, \cdots, f^{p-1}x,\ x,\ fx,\ \cdots,\ f^{p-1}x,\ x,\ fx,\ \cdots,\ f^{p-1}x,\ x, \cdots$$

Since by construction

$$\begin{Bmatrix} fx_n = x_{n+1} & \text{when } \ell \neq p-1 \\ d(fx_n, x_{n+1}) < \beta & \text{when } \ell = p-1 \end{Bmatrix}$$

we see that the sequence $(x_n)_{n=-\infty}^{+\infty}$ is a $(\delta_k)_{k=1}^{+\infty}$-pseudo orbit. We can therefore apply Theorem 5.1 to get a shadowing point $z\in M$ which, in particular, satisfies $d(z,x) < \alpha\epsilon_{k+1} < \alpha$.

Step 2. Notice that not only is z a shadowing point for $(x_n)_{n=-\infty}^{+\infty}$ but also $f^p z$ (since $x_{n+p}=x_n$, by construction). By Lemma 5.3 such shadowing points are unique. Therefore, we can conclude that $f^p z = z$. □

Remark. We have *not* shown that the periodic point lies in the Pesin set (or, equivalently, that it is a hyperbolic periodic point). However, this can be done with a little more work. The key point is that z always stays close to Λ in its orbit and so picks up enough 'hyperbolicity' that it must be in Λ itself. (The appropriate result to use here is 'Alexseev's Theorem'; cf.[Kat-Men].)

5.5. An application of the closing lemma.

The closing lemma (Theorem 5.2) is a very useful device for understanding properties of $f\colon M \longrightarrow M$. In the next chapter this will be explored more fully. However, to set the stage we shall conclude this chapter with a simple application which serves to illustrate the basic philosophy.

Proposition 5.1 If $f\colon M \longrightarrow M$ has a hyperbolic measure then there exists at least one periodic point (i.e. $\mathcal{M}^*_{\text{erg}} \neq \emptyset \Rightarrow \exists f^p x = x \in M$) .

Proof. To clarify the ideas in this simple proof it is useful to break it down into four steps.

Step 1. Choose any hyperbolic measure $m \in \mathcal{M}^*_{\text{erg}}$ and let $\Lambda = \bigcup_{k=1}^{+\infty} \Lambda_k$ be the associated Pesin set. Since $m(\Lambda) = 1$ (by Proposition 4.2 in Chapter 4) we must have $m(\Lambda_k) > 0$, for k sufficiently large.

Step 2. Using the closing lemma (Theorem 5.2) we can choose $\beta > 0$ corresponding to k (and any choice of $\eta > 0$).

We can then cover the compact set Λ_k by a finite number of open balls of radius $\beta/2$. Since $m(\Lambda_k) > 0$ at least one such open ball

$$B_{\Lambda_k}(x, \beta/2) = \{y \in \Lambda_k \mid d(x,y) < \beta/2\},$$

say, must have positive measure, i.e. $m\Big(B_{\Lambda_k}(x, \beta/2)\Big) > 0$.

Step 3. Since $m\Big(B_{\Lambda_k}(x,\beta/2)\Big) > 0$ we can apply the Poincaré recurrence theorem (Theorem 1.1) to see that almost all points in $B_{\Lambda_k}(x,\beta/2)$ return to $B_{\Lambda_k}(x,\beta/2)$ under iteration. In particular, we can find at least one such point $y \in B(x,\beta/2)$, say, and $p \geq 1$ such that $f^p y \in B_{\Lambda_k}(x, \beta/2)$.

In particular, this point satisfies $y, f^p y \in \Lambda_k$ and $d(y,f^p y) \leq d(y,x) + d(x,f^p y) \leq \beta/2+\beta/2 = \beta$.

Step 4. By applying the closing lemma (Theorem 5.2) we get that there exists a periodic point (of period p and close to the point y).

□

Since for surfaces with non-zero entropy our hypotheses are automatically true, we have the following elegant result of Katok.

Corollary 5.1.1 If $f\colon M \longrightarrow M$ is a $C^{1+\alpha}$ diffeomorphism of a compact surface with $h_{top}(f) > 0$ then there exists a periodic point for f.

Remarks. (i) The use of measures in the proof, but their absence in the hypothesis and the conclusion, is very characteristic of this approach. The main rôle of the measure is to introduce the necessary recurrence (through the Poincaré recurrence theorem) .

(ii) There are three assumptions in the corollary all of which are important. This is illustrated by the following simple examples.

(a) A counter-example with $h_{top}(f)=0$. Consider the diffeomorphism $f\colon \mathbf{R}^2/\mathbf{Z}^2 \longrightarrow \mathbf{R}^2/\mathbf{Z}^2$, defined by $f(x_1,x_2)+\mathbf{Z}^2 = (x_1+\alpha_1,x_2+\alpha_2)+\mathbf{Z}^2$ where $\alpha_1,\alpha_2 \in \mathbf{R}$ are irrational. This is a smooth surface diffeomorphism with no periodic points, but $h_{top}(f)=0$.

(b) A counter-example with $\dim M=3$. Choose any smooth surface diffeomorphism $g\colon N \longrightarrow N$ satisfying $h_{top}(g)>0$ then let $M=N\times S^1$ and define a diffeomorphism $f\colon M \longrightarrow M$ by $f(x,z)=(g(x), e^{2\pi i\alpha}z)$, where α is irrational. This is a smooth diffeomorphism with $h_{top}(f)=h_{top}(g)>0$ with no periodic points, but M is not a surface.

(c) A counter-example with $f\colon M \longrightarrow M$ continuous. There exist examples of homeomorphisms $f\colon \mathbf{R}^2/\mathbf{Z}^2 \longrightarrow \mathbf{R}^2/\mathbf{Z}^2$ with $h_{top}(f)>0$ which are even minimal (i.e. contain no closed f-invariant subsets) and so in particular no periodic points; see [Rees]. Rees' examples can be made Hölder continuous, but apparently not differentiable.

Notes

Liapunov neighborhoods were introduced in Pesin's 1976 articles and play a fundamental role in Katok's treatment of Pesin theory in his 1980 article [Katok]. The treatment we give is modified from Newhouse's later 1989 article [Newhouse_1].

The closing lemma occurs in Katok's article [Katok]. The statement of the slightly more general shadowing lemma was told to me by Newhouse (Katok also told me that a similar result will occur in his notes with Mendoza).

Our main application of these results to the existence of periodic points is taken directly from Katok's article.

Chapter 6.

The structure of 'chaotic' diffeomorphisms

In this chapter we shall use techniques developed in the previous lectures to understand the dynamical behavior of $C^{1+\alpha}$ diffeomorphisms for which there exist hyperbolic measures i.e. $\mathcal{M}^*_{erg} \neq \emptyset$ (including our main example of a surface diffeomorphism $f: M \longrightarrow M$ with $h_{top}(f) > 0$).

In this chapter we consider the following basic questions:

Question 1. Where in M are the periodic points? How many are there?

Question 2. What causes the complicated behaviour when $\mathcal{M}^*_{erg} \neq \emptyset$?

Question 3. How does $h_{top}: \mathrm{Diff}^{1+\alpha}(M) \longrightarrow \mathbb{R}^+$ vary?

(where $\mathrm{Diff}^{1+\alpha}(M)$ denotes the space of all $C^{1+\alpha}$ diffeomorphisms of M; see Appendix B.)

Our analysis will rest on: (i) the closing lemma (Proposition 5.1); and (ii) simple recurrence behaviour and Poincaré's Lemma (Theorem 1.1).

The first problem has clear similarities with the uniformly hyperbolic case. The second question leads us to consider the rôle of homoclinic points and horse-shoes in creating entropy. It then transpires that these simple examples of horse-shoes actually occur and carry most of the entropy of the diffeomorphism $f: M \longrightarrow M$. This gives some 'stability' to the entropy and addresses the third question.

6.1 The distribution of periodic points.

For any hyperbolic measure $m \in \mathcal{M}^*_{erg}$ we recall that the support $\mathrm{supp}(m)$

of this measure is defined to be the *smallest* closed set C satisfying $m(C) = 1$ (i.e. in some sense the measure m 'lives' on this set supp(m)).

Let $\text{Per}(f) = \{x \in M \mid f^p x = x, \text{ some } p \geq 1\}$ denote the set of *all* periodic points for $f: M \longrightarrow M$. We then have the following result about $\text{supp}(m) \subset M$.

Proposition 6.1 For any hyperbolic measure $m \in \mathcal{M}^*_{\text{erg}}$ the support of this measure is contained in the closure of the periodic orbits, i.e. $\text{supp}(m) \subseteq \text{cl}\big(\text{Per}(f)\big)$.

The proof is a variation of that for Proposition 5.1. We postpone it to the end of this chapter.

We illustrate this result with a few examples.

Examples (i) Let $M = \mathbb{R}^2/\mathbb{Z}^2$ and consider the diffeomorphism $f(x_1,x_2)+\mathbb{Z}^2 = (2x_1+x_2, x_1+x_2)+\mathbb{Z}^2$. The Lebesgue-Haar measure m on M has already been shown to be hyperbolic (as in fact is any ergodic measure). We can easily see that $\text{supp}(m) = M$, since every open ball in M has positive measure (with respect to m). Furthermore, one can easily see that $\text{cl}\big(\text{Per}(f)\big) = M$, as follows. Every point in the set

$$P_q = \{(\tfrac{p_1}{q}, \tfrac{p_2}{q}) \mid 0 \leq p_1, p_2 \leq q-1\}$$

must be a periodic point for f (since for each $i \geq 1$ we have

$$f^i(\tfrac{p_1}{q}, \tfrac{p_2}{q}) = (\tfrac{p_1'}{q}, \tfrac{p_2'}{q}),$$

i.e. the action of f^i simply permutes the finite set P_q, so when i is the length of a cycle the corresponding point is fixed). Furthermore, $\bigcup_q P_q \subset \text{Per}(f)$ is dense in M (since it exhausts all points with rational co-ordinates). Thus, in this case we have an equality $\text{supp}(m) = \text{cl}\big(\text{Per}(f)\big)$.

(ii) Let $M = \mathbb{R}^2/\mathbb{Z}^2$ and again let $f(x_1,x_2) + \mathbb{Z}^2 = (2x_1+x_2, x_1+x_2)+\mathbb{Z}^2$. This time we take the hyperbolic measure m to be the Dirac measure supported on the fixed point (0,0). As we observed before $\text{cl}(\text{Per}(f)) = M$, but in this case we have $\text{supp}(m) = (0,0)$. Thus, in this case we have a strict inclusion $\text{supp}(m) \subset \text{cl}\big(\text{Per}(f)\big)$.

(iii) Recall the standard Smale horse-shoe example. In this case there is an invariant Cantor set C which has a natural identification $C \equiv \prod_{-\infty}^{+\infty}\{0,1\}$. If we consider the invariant measure which corresponds to the Bernoulli measure $m=(\frac{1}{2},\frac{1}{2})^{\mathbf{Z}}$ then it is easy to see that $\mathrm{supp}\{(\frac{1}{2},\frac{1}{2})^{\mathbf{Z}}\} = C$. Furthermore, the periodic points for f can be identified in terms of periodic sequences in $\prod_{-\infty}^{+\infty}\{0,1\}$ *plus* the two additional fixed points which we denote by x_1,x_2. In particular, we have $\mathrm{cl}(\mathrm{Per}(f)) = C\cup\{x_1,x_2\}$ (which is precisely the non-wandering set for $f\colon M\longrightarrow M$, since this is an example of an Axiom A diffeomorphism) .

Remark. We recall that we showed earlier (in the remark in part (b) of the interlude) that for any $m\in\mathcal{M}_{\mathrm{inv}}$, $\mathrm{supp}(m)\subseteq\Omega=\{x\in M\mid \forall\text{neighborhoods } U\ni x, \exists n_i\to+\infty \text{ with } f^{n_i}U\cap U\neq\phi\}$, i.e. Ω is the non-wandering set. For Axiom A systems it is a basic assumption that $\Omega=\mathrm{cl}(\mathrm{Per}(f))$ (cf. Part (b) of the Interlude). Thus Proposition 6.1 can be viewed as a generalization of that special case.

6.2 The number of periodic points.

In the previous chapter we showed that whenever we have a hyperbolic measure then periodic points always exist (using the closing lemma and Poincaré recurrence). We now go further and and show that there are infinitely many periodic points (with asymptotic lower bounds). The proof refines the techniques used in the preceding section.

In particular, if $\mathrm{Per}_n(f) = \{x\in M \mid f^n x=x, \text{ some } n\geq 1\}$ and we let $|\mathrm{Per}_n f| = \mathrm{Card}\{\mathrm{Per}_n f\}$ be the cardinality of this set then we have:

Proposition 6.2 Whenever $m\in\mathcal{M}^*_{\mathrm{erg}}$ we have

$$\overline{\lim_{n\to+\infty}}\frac{1}{n}\log|\mathrm{Per}_n f| \geq h_{\mathrm{meas}}(m) > 0$$

We shall sketch the proof of this result later in the chapter.

For surfaces we have the following simple corollary.

Corollary 6.2.1 For *any* $C^{1+\alpha}$ surface diffeomorphism $f\colon M\longrightarrow M$ we have

$$\overline{\lim_{n\to+\infty}}\ \frac{1}{n}\log|\mathrm{Per}_n(f)| \geq h_{\mathrm{top}}(f) \tag{6.1}$$

Proof of Corollary 6.2.1. If $h_{\text{top}}(f) = 0$ then the result is trivial, therefore we can assume that $h_{\text{top}}(f)>0$. By the variational principle (Proposition 3.1) we can choose an ergodic measure $m\in\mathcal{M}_{\text{erg}}$ such that $h_{\text{meas}}(m) \geq h_{\text{top}}(f)-\epsilon>0$, for any given $\epsilon>0$ (see Chapter 3). By Theorem 3.2, in Chapter 3, any such measure m must be hyperbolic, i.e. $m\in\mathcal{M}^*_{\text{erg}}$. By Proposition 6.2 we get

$$\overline{\lim_{n\to+\infty}} \frac{1}{n}\log \text{Per}_n f \geq h_{\text{meas}}(m) \geq h_{\text{top}}(f)-\epsilon.$$

Finally, since $\epsilon>0$ can be arbitrarily small, the result follows. □

We call a diffeomorphism $f: M\longrightarrow M$ *expansive* if $\exists\epsilon>0$ such that whenever $x,y\in M$ with $d(f^n x,f^n y)<\epsilon$, $\forall n\in\mathbb{Z}$, then $x=y$. As an example of an expansive map, consider $M=\mathbb{R}^2/\mathbb{Z}^2$ and $f: M\longrightarrow M$ the diffeomorphism defined by $f(x_1,x_2)+\mathbb{Z}^2= (2x_1+x_2,\ x_1+x_2)+\mathbb{Z}^2$; cf. [Bowen].

If we assume that f is expansive then the set $\text{Per}_n(f)$ of periodic points of period n is always finite and we get the opposite inequality to (6.1)

$$\overline{\lim_{n\to+\infty}} \frac{1}{n} \log \text{Per}_n(f) \leq h_{\text{top}}(f)$$

(see [Walters], p.203). Thus, for expansive $C^{1+\alpha}$ surface diffeomorphisms:

$$h_{\text{top}}(f) = \overline{\lim_{n\to+\infty}} \frac{1}{n} \log \text{Per}_n(f).$$

Without the expansive condition we could have a strict inequality. For example, if f is the identity map the entropy is always zero but *every* point is a fixed point.

Examples. (i) Let $M=\mathbb{R}^2/\mathbb{Z}^2$ and let $f: M\longrightarrow M$ be the diffeomorphism defined by $f(x_1,x_2)+\mathbb{Z}^2 = (2x_1+x_2,x_1+x_2)$. As we have seen before $h_{\text{top}}(f) = \log(\frac{3+\sqrt{5}}{2})$. Since this is an example of an Axiom A diffeomorphism, we recall that we actually have a limit

$$\lim_{n\to+\infty} \frac{1}{n}\log \text{Per}_n(f) = \log\left(\frac{3+\sqrt{5}}{2}\right).$$

(ii) (Horse-shoe examples) Consider the standard Smale horse-shoe *or* one of the non-uniformly hyperbolic variants. It is well known that $h_{\text{top}}(f) = \log 2$ and the periodic points can be explicitly counted using the coding by sequences (i.e. $\mathrm{Per}_n(f) = 2^n+2$, where we have to remember to count the two additional fixed points). In particular, we get the limit

$$\lim_{n\to+\infty} \frac{1}{n} \log \mathrm{Per}_n(f) = \log 2.$$

6.3. Homoclinic points.

Assume $f^p x = x \in \Lambda$ is a periodic point. We call this fixed point *hyperbolic* if none of the eigenvalues of $D_x f$ has modulus unity.

Definition. We define the ***stable (unstable) manifold*** for x by

$$W_x^s = \{y \in M \mid d(f^n x, f^n y) \longrightarrow 0, \text{ as } n \longrightarrow +\infty\}$$

$$(W_x^u = \{y \in M \mid d(f^{-n} x, f^{-n} y) \longrightarrow 0, \text{ as } n \longrightarrow +\infty\})$$

We shall denote $n = dim E_x^u$, $\ell = \dim E_x^s$. The following is a basic lemma in dynamic systems.

Lemma 6.1 For a hyperbolic periodic point $x \in M$ the stable and unstable manifolds W_x^s, W_x^u are immersed C^1 submanifolds of dimensions ℓ and n, respectively, with tangent spaces $T_x W_x^s = E_x^s$, $T_x W_x^u = E_x^u$ (where E_x^u and E_x^s correspond to the eigenspaces for $D_x f^p$ with eigenvalues greater than unity and less than unity, respectively; see [Shub], for a proof).

Definition. We call any point $y \in W_x^s \cap W_x^u$, $y \neq x$, a ***(transverse) homoclinic point*** if the stable and unstable manifolds W_x^s, W_x^u meet transversely at y (that is, $T_y W_x^s \neq T_y W_x^u$).

This definition is illustrated in Figure 23

Remark. The importance of homoclinic points is that they are a source of very rich and complicated behavior. This will become more apparent in the next section.

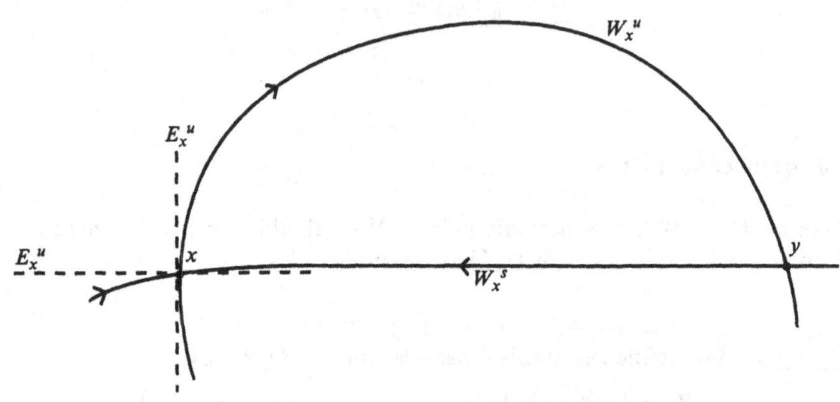

Figure 23: A transverse homoclinic point

The results we have stated before about periodic points remain true when we consider only periodic points which correspond to transverse homoclinic points. To illustrate this we mention the following result.

Proposition 6.3 For any $C^{1+\alpha}$ diffeomorphism $f: M \longrightarrow M$, on a compact manifold M, for which there exists a hyperbolic measure $m \in \mathcal{M}^*_{\text{erg}}$ which is *not* supported on a single closed orbit then there exists a (transverse) homoclinic point.

The proof is a variation on the existence result (Proposition 5.1) for periodic points given at the end of the last chapter. We sketch the proof later in this section. As usual, we have a simple corollary for surfaces.

Corollary 6.3.1 For a $C^{1+\alpha}$ diffeomorphism $f: M \longrightarrow M$, on a compact surface M, with $h_{top}(f)>0$ there exists a (transverse) homoclinic point.

Proof of Corollary 6.3.1. By the variational principle (Proposition 3.1) we can choose $m \in \mathcal{M}_{erg}$ with $h_{meas}(m) \geq h_{top}(f) - \epsilon > 0$, for any sufficiently small $\epsilon > 0$. We then observe:

(i) the measure m is hyperbolic, i.e. $m \in \mathcal{M}^*_{erg}$ (by Theorem 3.2);

(ii) m is *not* supported on a periodic orbit, since otherwise we would have $h_{meas}(m) = 0$.

Thus by Proposition 6.3 there exists a (transverse) homoclinic point. □

Let $\mathrm{Per}^h(f)$ be the set of periodic points $f^n x = x$ (for some $n \geq 1$) which have associated to them (transverse) homoclinic points, then we have a stronger result than Proposition 6.3.

Proposition 6.4 For $f: M \longrightarrow M$ a $C^{1+\alpha}$ diffeomorphism, $m \in \mathcal{M}^*_{erg}$ then $\mathrm{supp}(m) \subseteq \mathrm{cl}(\mathrm{Per}^h(f))$

The proof of Proposition 6.4 is sketched later in the chapter.

6.4. Generalized Smale horse-shoes.

To understand the importance of transverse homoclinic points we must first explain the kind of simple dynamical behavior they generate. We discussed earlier the standard (uniformly hyperbolic) Smale horse-shoe, where for a 'rectangle' $R \subseteq M$ we get the picture shown in Figure 24.

A *generalized Smale horse-shoe* refers to any similar situation for some 'rectangle' in M. A typical case is illustrated in Figure 25.

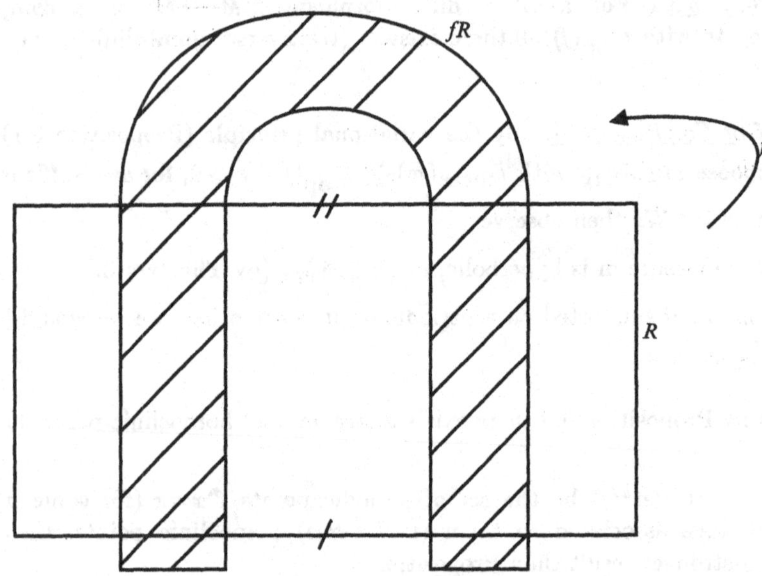

Figure 24: Standard Smale horse-shoe

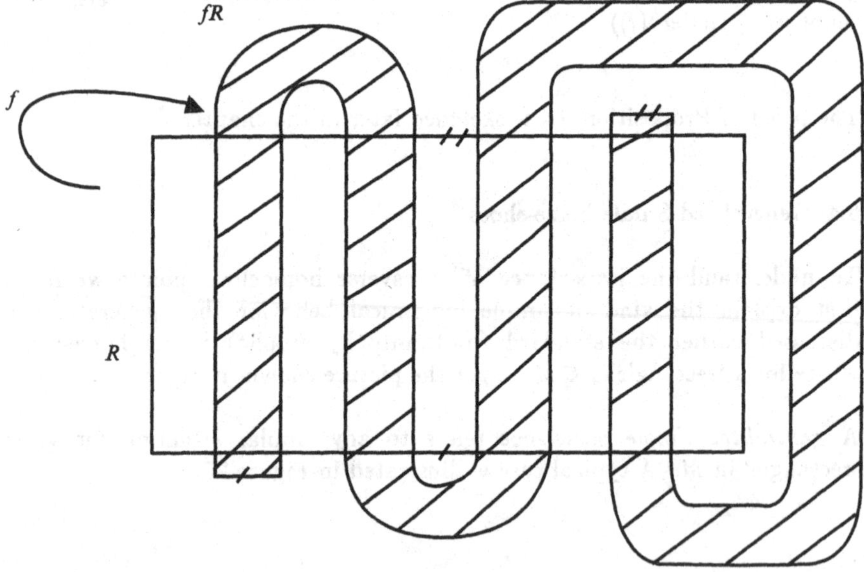

Figure 25: A generalized Smale horse-shoe

Usually horse-shoes are exhibited as *examples* of systems which demonstrate complicated dynamical behavior. In fact, we shall now see that they are implicit in the dynamics *whenever* we have transverse homoclinic points. This relationship between homoclinic points and generalized horse-shoes is given by the following standard theorem in dynamical systems.

Lemma 6.2 (Smale homoclinic theorem). If $f: M \longrightarrow M$ is a C^1 diffeomorphism, of a compact manifold M, with a (transverse) homoclinic point, then there exists a (generalized) horse-shoe in M for either f or some iterate f^n $(n > 0)$.

(see [Newhouse$_3$], p.18 or [Nitecki], p.154.)

Remark. The idea of proof, and the appropriate choice of the region R, is illustrated by Figure 26.

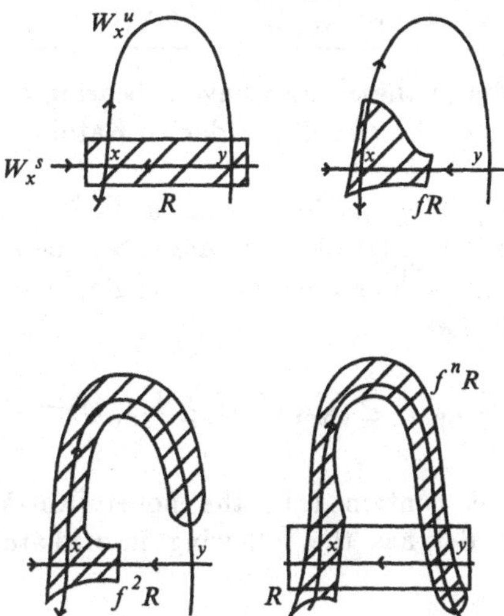

Figure 26: Smale homoclinic theorem

Combining Proposition 6.3 with the above lemma we find that horse-shoes are *always* present when we have 'chaotic' behavior (characterized by the existence of hyperbolic measures). We summarize these in the following additional corollaries to this proposition.

Corollary 6.3.2 If $f\colon M \longrightarrow M$ is a $C^{1+\alpha}$ diffeomorphism and there exist hyperbolic measures which are *not* supported on a single closed orbit then there exists a generalized horse-shoe.

As usual, we can simplify the hypothesis for surfaces to

Corollary 6.3.3 If $f\colon M \longrightarrow M$ is a $C^{1+\alpha}$ diffeomorphism with $\dim M=2$ and $h_{top}(f)>0$ then there exists a generalized horse-shoe.

6.5. Entropy stability.

We can associate to each given $C^{1+\alpha}$ diffeomorphism $f\colon M \longrightarrow M$ its topological entropy $h_{top}(f) \in \mathbb{R}^+$. In particular, if we vary the diffeomorphism f we can define a map $h_{top}\colon \mathrm{Diff}^{1+\alpha}(M) \longrightarrow \mathbb{R}^+$ by $f \mapsto h_{top}(f)$. This introduces the following immediate question.

Problem. What happens to $h_{top}(f)$ as we vary f?

The preceding results in these notes have a bearing on this problem. The following sharpening of these results is due to Katok.

Lemma 6.3 (Katok). Let $f\colon M \longrightarrow M$ be a $C^{1+\alpha}$ diffeomorphism of a compact surface with $h_{top}(f)>0$. For any $\epsilon>0$ there exists a generalized horse-shoe (where $R \subseteq M$ denotes the rectangle, and $\Lambda = \bigcap_{n \in \mathbb{Z}} f^{-n}R$ the recurrent part) such that:

$$h_{top}(f) \geq h_{top}(f|\Lambda) \geq h_{top}(f) - \epsilon$$

The proof is contained in the notes [Kat-Men] of Katok and Mendoza. Lemma 6.3 has the following immediate application to the above problem.

Proposition 6.5 Let $f\colon M \longrightarrow M$ be a $C^{1+\alpha}$ diffeomorphism of a compact surface with $h_{top}(f)>0$ then $h_{top}\colon \mathrm{Diff}^{1+\alpha}(M) \longrightarrow \mathbb{R}^+$ is lower semi-continuous. Moreover, given $\epsilon>0$ we have that for any $g \in \mathrm{Diff}^{1+\alpha}(M)$ sufficiently close to f (in the C^1 topology) $h_{top}(g) \geq h_{top}(f) - \epsilon$.

Proof. By Lemma 6.3 we know that there exists a generalized horse-shoe for $f: M \longrightarrow M$ such that $f: \Lambda \longrightarrow \Lambda$ satisfies $h_{top}(f|\Lambda) \geq h_{top}(f) - \epsilon$. Under small C^1 perturbations $f \longrightarrow g$ this horse-shoe will be persistent, i.e. g will have a 'topologically similar', or topologically conjugate, horse-shoe $g: \Lambda' \longrightarrow \Lambda'$. We need the C^1 hypothesis to ensure that for a rectangle R (for the generalized horse-shoe associated to f) the images $f(R)$ and $g(R)$ traverse R in the same configuration.

Since the topological entropy h_{top} is a conjugacy invariant (by Lemma 3.2) we deduce that $h_{top}(f|\Lambda) = h_{top}(g|\Lambda')$. From the definitions we see that $h_{top}(g) \geq h_{top}(g|\Lambda')$. Finally, by comparing the inequalities we see that $h_{top}(g) \geq h_{top}(g|\Lambda') = h_{top}(f|\Lambda) \geq h_{top}(f) - \epsilon$. □

Remarks. (i) Yomdin has shown that for manifolds of arbitrary dimensions the map $h_{top}: \mathrm{Diff}^\infty(M) \longrightarrow \mathbb{R}^+$ on C^∞ diffeomorphisms is upper semi-continuous (in the C^∞ topology) [Yomdin]. We shall consider these results further in Section 6.6. Combining Yomdin's result with the above proposition we get the following continuity result: *If M is a compact surface then* $h_{top}: \mathrm{Diff}^\infty(M) \longrightarrow \mathbb{R}^+$ *is continuous.*

(ii) In contrast to this continuity result for smooth diffeomorphisms of surfaces we observe that:

(a) the lower semi-continuity of h_{top} tends to fail in higher dimensions;

(b) the upper semi-continuity of $h_{top}: \mathrm{Diff}^k(M) \longrightarrow \mathbb{R}^+$, fails whenever k is finite. We shall consider specific examples in Section 6.7.

(iii) Newhouse has developed Yomdin's techniques to show that for a fixed C^∞ map $f: M \longrightarrow M$ the map $h_{meas}: \mathcal{M}_{inv} \longrightarrow \mathbb{R}^+$ is upper semi-continuous with respect to weak* topology [Newhouse$_1$]. In particular, since every upper semi-continuous map on a compact space realizes its supremum this implies the existence of measures of maximal entropy. We will not present a proof of this result. Instead, we shall prove an elementary analogous result for the measure theoretic entropy (relative to a fixed diffeomorphism f).

Proposition 6.6 Let $m \in \mathcal{M}$ be a smooth probability measure on a compact surface M and let $\mathrm{Diff}^1_m(M) \subset \mathrm{Diff}^1(M)$ be the subspace of diffeomorphisms for which m is an ergodic measure. Then the map $\mathrm{Diff}^1_m(M) \longrightarrow \mathbf{R}^+$ defined by $f \mapsto h_{\mathrm{meas}}(m,f)$ is upper semi-continuous, i.e. given $\epsilon > 0$, for any $g \in \mathrm{Diff}^1_m(M)$ sufficiently close to f we have $h_{\mathrm{meas}}(m,g) \leq h_{\mathrm{meas}}(m,f)+\epsilon$.

Proof. We saw from Propositions 2.1 and 2.2 that if $m \in \mathcal{M}_{\mathrm{erg}}$ then the limit $\lim_{n\to+\infty} \frac{1}{n}\log \|D_x f^n\| = \lambda$ exists almost everywhere and equals

$$\lambda = \inf\left\{\frac{1}{n}\int \log\|D_x f^n\| dm(x) \ \middle|\ n \geq 1\right\} \ \text{a.a.}(m) \ x \in M$$

It is easy to see that for a fixed measure m the maps $f \longrightarrow \frac{1}{n}\int\log\|D_x f^n\|dm(x)$, $n\geq 1$, have a continuous dependence on $f \in \mathrm{Diff}^1(M)$. Thus it follows that $f \mapsto \lambda = \lambda(f)$ is upper semi-continuous. If M is a surface with $h_{\mathrm{meas}}(m,f) > 0$ then by Theorem 3.2 (ii) we see that $\lambda_1 > 0$ is the only positive Liapunov exponent. From the proof of the Oseledec theorem (Theorem 2.1) we see that $\lambda_1 = \lambda(f)$.

Finally, we need to relate λ to the entropy $h_{\mathrm{meas}}(m,f)$. To do this we recall Pesin's result (in Section 3.2) that if $m \in \mathcal{M}_{\mathrm{erg}}$ is absolutely continuous then $h_{\mathrm{meas}}(m,f) = \log\lambda_1$. □

6.6 Entropy, volume growth and Yomdin's inequality.

We now want to give a brief over-view of Yomdin's estimates on volume growth (based on the Bourbaki seminar [Gromov]). This material deviates somewhat from the theme of these notes in that it does *not* depend on ideas from Pesin theory.

Given a C^1 diffeomorphism $f: M \longrightarrow M$ of a compact manifold and a ℓ-dimensional submanifold $N \subset M$ we want to consider the images $f^k N \subset M$, under the kth iterate of the diffeomorphism f. We denote by $\mathrm{Vol}(f^k N)$ the ℓ-dimensional volume of the image $f^k N$. More generally, we could consider smooth non-invertible maps $f: M \longrightarrow M$ if we adopt the convention of counting the volumes of self-intersections according to multiplicity.

We shall begin with some simple considerations. It is easy to see that

$$\log(\mathrm{Vol}(f^k N)) \leq \ell \log(\|Df^k\|) + \log(\mathrm{Vol}(N))$$

$$\leq \ell\, k \log(\|Df\|) + \log(\mathrm{Vol}(N)). \qquad (6.2)$$

If we introduce the convenient notation:

(i) $\mathrm{Vol}_f(N) = \overline{\lim}_{k\to\infty} \mathrm{Vol}(f^k N)^{1/k}$; and

(ii) $\lambda(f) = \overline{\lim}_{k\to\infty} \|Df^k\|^{1/k}$

then dividing by k and letting $k\to\infty$ in (6.2) we get that

$$\log(\mathrm{Vol}_f(N)) \leq \ell \log(\lambda(f)) \leq \ell \log(\|Df\|) \qquad (6.3)$$

Yomdin's observation was that the simple inequality (6.3) may be significantly improved if we assume that $f: M\to M$ has higher differentiability.

Theorem 6.1 (Yomdin). If $f: M\to M$ is a C^r diffeomorphism $(1 \leq r < \infty)$ and $N\subset M$ a C^r submanifold, then we have the inequality

$$\log(\mathrm{Vol}_f(N)) \leq h_{top}(f) + \tfrac{\ell}{r} \log(\lambda(f)) \qquad (6.4)$$

(and for $r = \infty$, $\log(\mathrm{Vol}_f(N)) \leq h_{top}(f)$)

We sketch part of the proof later in this chapter. We first consider two applications of this theorem.

I. Application to homology (The Shub Entropy Conjecture). The map $f: M\to M$ induces an action on the real homology groups $H_*(M,\mathbb{R})$, which we denote by $f_*: H_*(M,\mathbb{R})\to H_*(M,\mathbb{R})$. The spaces $H_*(M,\mathbb{R})$ are merely finite-dimensional Euclidean spaces and f_* corresponds to a linear map (represented as a finite matrix, with integer coefficients, by the universal coefficient theorem). Therefore, we can denote the spectral radius by $\rho(f_*)$. As we have already remarked in Section 3.4, we always have the inequality $\log\rho(f_1) \leq h_{top}(f)$ for the action on the *first* homology group. More generally, we have the following well-known conjecture of Shub.

Shub entropy conjecture. For any C^1 map $f: M\to M$, on a compact manifold M, we always have the inequality $\log\rho(f_*) \leq h_{top}(f)$. As we observe below, Yomdin's theorem shows immediately that this conjecture is true for C^∞ diffeomorphisms.

Corollary 6.1.1 For any C^∞ diffeomorphism $f: M \longrightarrow M$ we have that $\log \rho(f_*) \leq h_{top}(f)$.

Remark. Notice that for the trivial map $f = id$ we have an equality since $h_{top}(f)=0$ and $\rho(f_*)=1$.

Proof of Corollary 6.1.1 (assuming Theorem 6.1). The corollary follows from the general principle that 'volume growth dominates homological growth'. We now try to be more specific. By duality, it suffices to consider $\rho(f_*)=\rho(f^*)$, where $f^*: H^*(M,\mathbb{R}) \longrightarrow H^*(M,\mathbb{R})$ denotes the induced action on (de Rham) cohomology (see Appendix B). We can define an inner product on the space forms by

$$<\omega,\omega'> = \int *(\omega \wedge *\omega') \, d(\mathrm{Vol}),$$

where $*$ is the standard operator on spaces of forms (see Appendix B). In particular, if ω has support $\mathrm{supp}(\omega)=N \subset M$ then we have

$$\|(f^n)^*\omega\| = \left(\int *(f^n\omega \wedge [*f^n\omega]) \, d(\mathrm{Vol})\right)^{1/2} \leq \|\omega\|_\infty \, \mathrm{Vol}(f^n N)$$

and by taking nth roots and letting $n \longrightarrow +\infty$ we have

$$\rho(f^*) \;=\; \lim_{n\to\infty} \|(f^n)^*\omega\|^{1/n} \leq \mathrm{Vol}_f(N) \leq \exp\big(h_{top}(f)\big)$$

(by Theorem 6.1) which completes the proof of the Corollary. □

Remark. Notice that in the Shub entropy conjecture *some* differentiability is required, indeed we can give a simple example of a C^0 map for which the corresponding statement of the conjecture would be false. Consider the continuous map $f: \hat{\mathbb{C}} \longrightarrow \hat{\mathbb{C}}$ on the Riemann sphere by

$$f(z) = \begin{cases} 2z^2 & \text{if } z \neq \infty \\ \infty & \text{if } z = \infty \end{cases}$$

It is easy to see that the second homology group $H_2(\hat{\mathbb{C}},\mathbb{Z}) \equiv \mathbb{Z}$ and that the induced action $f_*: H_2(\hat{\mathbb{C}},\mathbb{Z}) \longrightarrow H_2(\hat{\mathbb{C}},\mathbb{Z})$ is given by $x \mapsto 2x$. In particular, we have that $\rho(f_*)=2$ and it is easy to see that this map has $h_{top}(f)=0$.

(NB. Since the non-wandering set consists of just the set of two points $\{0,\infty\}$, all measures in $\mathcal{M}_{\text{inv}}$ must be supported on these two points; see Remark in part (b) of Interlude. However, all such measures clearly have zero entropy, and the conclusion follows from Proposition 3.1 (ii)). However, observe that the map f fails to be differentiable at ∞.

We now want to make a few brief comments on the ingredients in Yomdin's proof (based on the survey [Gromov]).

First note that we can use the Whitney embedding theorem to assume that the manifold M is embedded in $\mathbb{R}^n$ (for some $n \geq 1$). We shall let $\|f\| = \max\{\|f\|_\infty, \|Df\|_\infty, \cdots, \|D^r f\|_\infty\}$, where $\| \ \|_\infty$ denotes the supremum norm on $C^0(M,\mathbb{R}^n)$, be the best bound on the first r derivatives. Secondly, instead of considering ℓ-dimensional C^r submanifolds $N \subseteq M$ it is more convenient to consider C^r maps $h\colon [0,1]^\ell \longrightarrow M \subset \mathbb{R}^n$. Given such a map h we shall denote its image by Y = image(h) and write $|Y| = \|h\| := \max\{\|h\|_\infty, \|Dh\|_\infty, \cdots, \|D^r h\|_\infty\}$.

A simple calculation using the chain rule shows that if $|Y| \leq 1$ then $|f(Y)| \leq Const.\|f\|$ (where the constant depends only on r, ℓ, and $\dim(M)$) and we can conclude that $f(Y)$ can be divided into a union of at most $u = [Const.\|f\|]+1$ pieces $Y_1, \cdots, Y_u$ with $|Y_i| \leq 1$. At the center of Yomdin's argument is the following more sophisticated variant of this estimate.

Estimate. If $|Y| \leq 1$, then for any unit cube $I \subset \mathbb{R}^n$ the intersection $f(Y) \cap I$ can be divided into $v = [Const.\|f\|]^{\ell/r}+1$ pieces $Y_1, \cdots, Y_v$ with $|Y_i| \leq 1$; cf. [Gromov].

The proof involves approximating f by polynomials in its Taylor series expansion and using algebraic properties on the associated varieties.

If we let $I_0, \cdots, I_{k-1} \subset \mathbb{R}^n$ be arbitrary unit cubes then by induction on k the set $f^k(Y) \cap I_0 \cap fI_1 \cap \cdots \cap f^{k-1} I_{k-1}$ can be divided into $w = \left([Const.\|f\|]^{\ell/r}+1\right)^k$ pieces $Y_1, \cdots, Y_w$ with $|Y_i| \leq 1$. Since $\|Dh_i\|_\infty \leq |Y_i| \leq 1$ we deduce that $\mathrm{Vol}(Y_i) \leq 1$ for $1 \leq i \leq w$ and conclude that

$$\mathrm{Vol}\left(f^k(Y) \cap I_0 \cap fI_1 \cap \cdots \cap f^{k-1} I_{k-1}\right) \leq \left([Const.\|f\|]^{\ell/r} + 1\right)^k \qquad (6.5)$$

Let $K \subset M$ be a (n,ϵ)-covering set with $\mathrm{Card}(K) = N(n,\epsilon)$, then by (6.5) we have that

$$\mathrm{Vol}(f^k(Y)) = \mathrm{Vol}\Big(\bigcup_{x\in K} f^k(Y)\cap D_\epsilon(x;d_k)\Big) \leq N(n,\epsilon)\left([Const.\|f\|]^{\ell/r}+1\right)^k \quad (6.6)$$

(provided $\epsilon \leq \frac{1}{2}$, since then $D_\epsilon(x;d_k) \subset I_0 \cap f I_1 \cap \cdots \cap f^{k-1} I_{k-1}$).

We can take logarithms of each side of (6.6), divide by k and let $k \longrightarrow \infty$,

$$\begin{aligned}
\log(\mathrm{Vol}_f(Y)) &= \overline{\lim_{k\to\infty}} \tfrac{1}{k} \log(\mathrm{Vol}(f^k Y)) \\
&\leq \overline{\lim_{k\to\infty}} \tfrac{1}{k} \log N(n,\epsilon) + \overline{\lim_{k\to\infty}} \log\left([Const.\|f\|]^{\ell/r}+1\right) \\
&= \overline{\lim_{k\to\infty}} \tfrac{1}{k} \log N(n,\epsilon) + \log(Const.) + \tfrac{\ell}{r}\|f\|
\end{aligned}$$

and letting $\epsilon \longrightarrow 0^+$ we get

$$\log(\mathrm{Vol}_f(Y)) \leq h_{\mathrm{top}}(f) + \log(Const.) + \tfrac{\ell}{r}\log(\|f\|) \quad (6.7)$$

To finish the proof of the theorem from this point we need only two simplifications.

(a) We are free to choose the scale of the co-ordinates in $\mathbf{R}^n$, and we observe that scaling has the most significant changes for the first derivative. In particular, we can choose a scale such that $\|Df\|_\infty/\|f\| \leq 1$ is arbitrarily close to unity.

(b) If we replace f by f^N, $N \geq 1$, in the above estimates and then divide thru by N then (6.7) is replaced by the identity

$$\log(\mathrm{Vol}_f(Y)) \leq h_{\mathrm{top}}(f) + \frac{\log(Const.)}{N} + \tfrac{\ell}{r}\log(\|f\|).$$

Thus, by letting $N \longrightarrow +\infty$ in (b) we get that

$$\log(\mathrm{Vol}_f(Y)) \leq h_{\mathrm{top}}(f) + \tfrac{\ell}{r}\log(\|f\|).$$

By applying (a) we get that

$$\log(\mathrm{Vol}_f(Y)) \leq h_{\mathrm{top}}(f) + \tfrac{\ell}{r}\log(\|Df\|_\infty).$$

To get the final form of (6.4) we need only replace f by the iterate f^n, divide both sides of the last inequlity by n, and then let $n \longrightarrow +\infty$.

II. Upper semi-continuity of topological entropy. In Chapter 3, we briefly mentioned Yomdin's result on the upper semi-continuity of topological entropy (in the C^∞ topology). The proof of this result is based on the techniques used in proving Theorem 6.1.

To make use of the C^r assumption, we introduce a C^r 'co-ordinate' function g: $[0,1]^\ell \longrightarrow M$ and define a *C^r (n,ϵ)-cover* to be a collection of C^r maps $h_1,\cdots,h_m$: $[0,1]^\ell \longrightarrow [0,1]^\ell$ such that

(a) $\bigcup_{i=1}^{m} \text{image}(h_i) = [0,1]^\ell$;

(b) $\|h_i\|$, $\|h_i \circ f^j \circ g\| \leq \epsilon$, for $1 \leq i \leq m$, and $0 \leq j \leq n-1$.

and let $N_{r,g}(n,\epsilon)$ denote the smallest number of elements required for such a cover. We can get rid of the 'co-ordinate functions' by defining

$$h_{r,\epsilon}^{(n)}(f) = \sup\{ \tfrac{1}{n} \log\Big((N_{r,g}(n,\epsilon)\Big) \mid \|g\| \leq 1\} \qquad (6.8)$$

and then we remove $\epsilon>0$ by taking the limit

$$h_r(f) = \overline{\lim_{\epsilon \to 0+}} \inf\{ h_{r,\epsilon}^{(n)}(f) \mid n \geq 1\}$$

In fact, it is even unnecessary to take the limit as $\epsilon \longrightarrow 0$ (from above), since the infimum can be shown to be constant, for ϵ sufficiently small. Therefore, the quantity $h_r(f)$ is semi-continuous as a function $f \in C^r(M,M)$.

By modifying the proof of the theorem it can be shown that

$$h_r(f) \leq h_{\text{top}}(f) + \tfrac{\ell}{r} \log(\|Df\|_\infty)$$

and, with a little more work, that $h_{\text{top}}(f) = \lim_{r \to \infty} h_r(f)$. The semi-continuity of $f \mapsto h_{\text{top}}(f)$ comes from the semi-continuity of $f \mapsto h_r(f)$.

For fuller details of I and II, the reader is referred to [Gromov].

6.7 Examples of discontinuity of entropy.

In Section 6.5, we made some observations about the semi-continuity and continuity properties of both topological and metric entropy. Now we want to put this into perspective by giving some simple (counter-) examples to continuity under weaker hypothesis.

In Section 6.5 we described Yomdin's result that the topological entropy map $h_{\text{top}}: C^\infty(M,M) \longrightarrow \mathbb{R}^+$ is upper semi-continuous (in the C^∞ topology). We can give two very simple examples to show that without the smoothness assumption the map $h_{\text{top}}: C^0(M,M) \longrightarrow \mathbb{R}^+$ may not be upper semi-continuous (in the C^0 topology).

Example 1 (Topological entropy and continuous maps). Let M be the two-dimensional sphere, which for simplicity we identify with the Riemann sphere $\hat{\mathbb{C}}$. Let $f_0: \hat{\mathbb{C}} \longrightarrow \hat{\mathbb{C}}$ be the continuous map (of degree 2) defined by extending the map

$$f_0: \begin{cases} \mathbb{C} \longrightarrow \mathbb{C} \\ z \longrightarrow z^2 \end{cases}$$

on the complex plane $\mathbb{C}$ by defining $f_0(\infty) = \infty$ (although, as we observed before, this fails to be differentiable at $z = \infty$). Since the restriction of this map to $S^1 = \{z \in \mathbb{C} \mid |z| = 1\}$ is the usual 'doubling map' on the unit circle, with topological entropy equal to log2, we conclude that $h_{\text{top}}(f_0) \geq \log 2$. (In fact, it is not difficult to verify that $h_{\text{top}}(f_0) = \log 2$.)

We can define a continuous family of maps $f_t \in C^0(M,M)$, $t \in [-\frac{1}{2},\frac{1}{2}]$, by

$$f_t(z) = \begin{cases} (1+t)\, z^2 & \text{if } z \neq \infty \\ \infty & \text{if } z = \infty \end{cases}$$

For each member of this family there are fixed points $0,\infty \in \hat{\mathbb{C}}$. If $t > 0$ then for any point $z \in \hat{\mathbb{C}} - \{0,\infty\}$ we have $f_t^{\,n} z \longrightarrow \infty$, and if $t < 0$ then for any point $z \in \hat{\mathbb{C}} - \{0,\infty\}$ we have $f_t^{\,n} z \longrightarrow 0$, as $n \longrightarrow +\infty$. In either of these two cases, any f_t-invariant probability measure μ, say, *must* be supported on the finite set $\{0,\infty\}$ and thus trivially satisfy $h_{\text{meas}}(\mu)=0$. It then follows from the variational principle (Proposition 3.1) that $h_{\text{top}}(f_t)=0$ for $t \in [-\frac{1}{2},0) \cup (0, \frac{1}{2}]$. In particular, the map is not upper semi-continuous at $t=0$.

Example 2 (Topological entropy and smooth maps). In Misiurewicz's original examples of discontinuities of entropy, he considered a family of C^∞ diffeomorphisms [Misiurewicz]. Specifically, let $\mathbb{T}^2 = \mathbb{R}^2/\mathbb{Z}^2$ be the two-dimensional torus and define a map $T: \mathbb{T}^2 \longrightarrow \mathbb{T}^2$ by $T((x_1,x_2)+\mathbb{Z}^2) = (2x_1+x_2, x_1+x_2)$. Let $N = \mathbb{T}^2 \times [0,1]/\sim$ be the mapping torus, where we identify $(x,1) \sim (Tx,0)$, and let $\phi_t: N \longrightarrow N$ be the suspended flow

$\phi_t(x,u)=(x,u+t)$ (subject to the identifications). Clearly, the time-one flow $\phi_{t=1}$: $N\longrightarrow N$ has non-zero topological entropy $h_{\text{top}}(\phi_{t=1})>0$.

We let F_n: $\mathbb{T}^1\longrightarrow\mathbb{T}^1$, $n\geq 1$, be the family of C^∞ maps of the circle $\mathbb{T}^1$ given by $F_n(x) = x+\frac{1}{n}x(1-x)$. In particular, we see that $0\equiv 1$ is a fixed point and for every $y\in(0,1)$ we have $F_n(y)\longrightarrow 0$, as $n\longrightarrow+\infty$. Furthermore, F_n converges to the identity in the C^∞ topology as $n\longrightarrow+\infty$.

We set $M=N\times\mathbb{T}^1$ and define f: $M\longrightarrow M$ by $f(w,x) = (\phi_{\sin(2\pi x)}w,x)$ and f_n: $M\longrightarrow M$ by $f_n(w,x) =(\phi_{\sin(2\pi x)}w, F_n(x))$, for $n\geq 1$, then clearly $f_n\longrightarrow f$ in the C^∞ topology as $n\longrightarrow+\infty$.

(I) For each $n\geq 1$ the non-wandering set of f_n consists of $\Omega=N\times\{0\}$ and f_n: $\Omega\longrightarrow\Omega$ is the identity. It therefore follows that $h_{\text{top}}(f_n)=0$, for $n\geq 1$.

(II) The submanifolds $N_x=N\times\{x\}$ are each f-invariant. Furthermore, if $x\neq 0,\frac{1}{2}$, then f: $N_x\longrightarrow N_x$ has non-zero topological entropy. In particular, this implies that the topological entropy on the whole space is non-zero i.e. $h_{\text{top}}(f)>0$.

In Proposition 6.6 we also observed that on the subspace $\text{Diff}^1_m(M) \subset \text{Diff}^1(M)$ of diffeomorphisms which preserve $m\in\mathcal{M}_{\text{erg}}$ the map $f\mapsto h_{\text{meas}}(m,f)$ is upper semi-continuous. We can now give a simple example to show that this map may not be lower semi-continuous.

<u>Example 3</u> (Measure theoretic entropy and smooth maps). Again let $\mathbb{T}^2=\mathbb{R}^2/\mathbb{Z}^2$ and define T:$\mathbb{T}^2\longrightarrow\mathbb{T}^2$ by $T((x_1,x_2)+\mathbb{Z}^2) = (2x_1+x_2,x_1+x_2)$. As before, let $N=\mathbb{T}^2\times[0,1]/\sim$ with the identification $(x,1)\sim(Tx,0)$, and ϕ_t: $N\longrightarrow N$ is the suspended flow $\phi_t(x,u)=(x,u+t)$. This is a uniformly hyperbolic flow with $TN=E_1\oplus E_2\oplus E_3$ being the hyperbolic splitting associated to the flow ϕ_t, which has exponents $\pm\log((3+\sqrt{5})/2)$, 0.

Finally, we define the four-dimensional manifold $M=N\times\mathbb{T}^1$ (where $\mathbb{T}^1\subset\mathbb{C}$ denotes the unit circle) and a C^∞ (even real analytic) family of diffeomorphisms f_α: $M\longrightarrow M$, given by

$$f_\alpha(z,w) = (\phi_{\Im(w)}(z),w\,e^{2\pi i\alpha}),\text{ where } 0\leq\alpha\leq 1.$$

Therefore, $f_\alpha{}^n(z,w) = (\phi_{t(n,\alpha,w)}(z), we^{2\pi in\alpha})$, where $t(n,\alpha,w) = [\sum_{k=0}^{n-1}\Im(we^{2\pi ik\alpha})]$, for all $(z,w)\in M$.

Notice there exists a natural splitting $TM=E_1\oplus E_2\oplus E_3\oplus E_4$ into Df_α-invariant sub-bundles, for each $0\leq\alpha\leq 1$.

It is easy to see that *each* of these diffeomorphisms preserves the normalized volume on M, which we shall denote by μ. Furthermore, it is easy to see that the Liapunov exponents associated to the bundles $E_3\oplus E_4$ are always zero. However, the ergodic properties of this measure and the remaining two Liapunov exponents depend on the value of α:

Case I. If $\alpha\in[0,1]-\mathbb{Q}$ is irrational, then $m\in\mathcal{M}_{erg}$. Since $t(n,\alpha,w) = o(n)$ (i.e. $t(n,\alpha,w)/n\longrightarrow 0$, as $n\longrightarrow+\infty$), this implies that $\log\|Df_\alpha{}^n|E_1\|$, $\log\|Df_\alpha{}^n|E_2\| = o(1)$ and therefore we see that the Liapunov exponents associated to these bundles are also zero. Hence, all four Liapunov exponents associated to m are zero and by the Pesin-Ruelle inequality (Theorem 3.1) we have that $h_{meas}(\mu) = 0+0+0+0 = 0$.

Case II. If $\alpha\in [0,1]\cap\mathbb{Q}$ is rational then, $m\notin\mathcal{M}_{erg}$. Consider, for example, the value $\alpha=0$, then

$$t(n,0,w) = [\sum_{k=0}^{n-1} \Im m(w)] = n\Im m(w),$$

for all $n\in\mathbb{Z}$ and $w\in K$, and that the submanifolds $N_{w_0}=N\times\{w_0\}$ with coefficients $w=w_0$ are each invariant. If we consider the restriction of $Df_\alpha{}^n$ to $E_1\oplus E_2$ we see that:

$$\|Df_\alpha{}^n|E_1\| = \log\left(\frac{3+\sqrt{5}}{2}\right)\, n\,\Im m(w)$$

$$\|Df_\alpha{}^n|E_2\| = -\log\left(\frac{3+\sqrt{5}}{2}\right)\, n\,\Im m(w)$$

and deduce that the two remaining Liapunov exponents are $\pm\log((3+\sqrt{5})/2)\,\Im m(w)$. (Clearly, these depend on which submanifold $N_w\subset M$ we start from, a property consistent with m *not* being ergodic.) By the general form of Pesin's theorem the entropy $h_{meas}(m)$ is equal to the integral of the positive Liapunov exponents. In this particular case this becomes

$$h_{meas}(m) = \int\left|\log\left(\frac{3+\sqrt{5}}{2}\right)\,\Im m(w)\right|\,dm = \tfrac{1}{2}\log\left(\frac{3+\sqrt{5}}{2}\right)$$

Comparing these two cases, we see that the measure theoretic entropy $\alpha\mapsto f_\alpha\mapsto h_{meas}(m)$ is identically zero on the irrationals, but non-zero on the rationals (in particular, at $\alpha=0$). Therefore, we conclude that, although this map is upper semi-continuous, it is not lower semi-continuous.

6.8 Proofs of propositions 6.1 and 6.2

In this section we shall return to some of the proofs omitted earlier in this chapter.

Proof of Proposition 6.1. The proof of Proposition 6.1 is just an improved version of the proof of the Proposition 5.1 in Section 5.5).

Step 1. To prove the result it suffices to show that for any $x \in \text{supp}(m)$, and any $\epsilon > 0$, there exists a periodic point $f^p z = z$ (for some $p \geq 1$) with $d(x,z) < \epsilon$. Since $x \in \text{supp}(m)$ we can immediately deduce $m(B_\Lambda(x,\epsilon)) > 0$, where we denote

$$B_\Lambda(x,\epsilon) = \{y \in \Lambda \mid d(y,x) < \epsilon\}.$$

Since we can write the Pesin set as $\Lambda = \bigcup_{k=1}^{+\infty} \Lambda_k$ we see that for sufficiently large $k \geq 1$ we have $m(B_{\Lambda_k}(x,\epsilon)) > 0$, where we denote

$$B_{\Lambda_k}(x,\epsilon) = \{y \in \Lambda_k \mid d(y,x) < \epsilon\}.$$

Step 2. For the choice of k above we choose the value $\beta = \beta(k, \epsilon/2)$ given by the closing lemma (Theorem 5.2), where we can also assume, without loss of generality, that $\beta < \epsilon$. Choose a cover for the set $B_{\Lambda_k}(x,\epsilon)$ using a finite number of $(\beta/2)$-discs (in Λ_k) and then choose an element of this cover $B_{\Lambda_k}(x_0, \beta/2)$, say, satisfying $m(B_{\Lambda_k}(x_0,\beta/2)) > 0$.

Step 3. By Poincaré recurrence almost all points in the set $B_{\Lambda_k}(x_0,\beta/2)$ return under iteration of f. In particular, we can choose one such point $y \in B_{\Lambda_k}(x_0,\beta/2)$ with $f^p y \in B_{\Lambda_k}(x_0,\beta/2)$, for some $p \geq 1$. Applying the closing lemma (Theorem 5.2) we have that there exists a periodic point $z = f^p z \in M$ with $d(z,y) < \beta/2$. Therefore, we can conclude that $d(z,x) < d(z,y) + d(y,x) < \beta/2 + \beta/2 = \beta < \epsilon$. This completes the proof. □

Proof of Proposition 6.2. Recall that the measure theoretic entropy was defined by

$$h_{\text{meas}}(m) = \lim_{\delta\to 1} \lim_{\epsilon\to 0} \lim_{n\to+\infty} \frac{1}{n} \log N(n,\epsilon;\delta),$$

where $N(n,\epsilon;\delta) = \min\{\text{Card}(K) \mid K \text{ is a } (n,\epsilon;\delta)\text{-covering set}\}$, and each $(n,\epsilon;\delta)$-covering set satisfies

$$m\Big(\bigcup_{x\in K} B_\epsilon(x,d_n)\Big) > \delta.$$

(We also recall the comment in Section 3.1, that these limits are independent of $\delta>0$.)

The idea of the proof of Proposition 6.2 is to approximate the sets K by periodic points using the closing lemma. (This idea is familiar from the Axiom A case.) Since the specific details in the proof are complicated we refer the reader to [Katok], pp.169-171, although we emphasize again that the basic idea is easy. □

Proof of Proposition 6.3. We shall follow the proof of Proposition 6.1, with some minor variations. Choose any point $x\in\text{supp}(m)$, then for any $\epsilon>0$ we have $m(B_\Lambda(x,\epsilon))>0$. We proceed as follows.

Step 1. Choose $k\geq 1$ sufficiently large that $m(B_{\Lambda_k}(x,\epsilon))>0$.

Step 2. For the choice of k above, choose two points $x_1,x_2\in B_{\Lambda_k}(x,\epsilon)$ satisfying:

(i) $x_1 \neq x_2$;

(ii) $\forall\delta > 0,\ m(B_{\Lambda_k}(x_1, \delta)),\ m\,(B_{\Lambda_k}(x_2, \delta)) > 0$;

(iii) the distance $d = d(x_1,x_2)$ is 'small' (in a sense to be clarified in Step 4).

The existence of such points requires the use of results from Chapter 4.

We can now choose smaller discs $B_1 \subseteq B_{\Lambda_k}(x_1,\frac{d}{3})$ and $B_2 \subseteq B_{\Lambda_k}(x_2,d/3)$ such that $m(B_1)$, $m(B_2)>0$ and $\text{diam}(B_1), \text{diam}(B_2) < \beta = \beta(k,d/3)$ (using the value $\beta > 0$ specified in the statement of the closing lemma, Theorem 5.2).

Step 3. By using Poincaré recurrence for the sets B_1, B_2 and applying the closing lemma there exist periodic points $f^{p_1}z_1=z_1$ and $f^{p_2}z_2=z_2$ (p_1, $p_2 \geq 1$) satisfying $d(z_2,x_2)<d/3$ and $d(z_1,x_1)<d/3$. This leads us to the following inequalities:

$$d(z_1, z_2) \leq d(z_1, x_1) + d(x_1, x_2) + d(x_2, z_2) \leq (\tfrac{5}{3})\, d$$

and

$$d(z_1, z_2) \geq d(x_1, x_2) - d(x_1, z_1) - d(x_2, z_2) \geq (\tfrac{1}{3})\, d$$

In particular, the second inequality tells us that $z_1 \neq z_2$.

Step 4. We now specify that d should be sufficiently 'small' that the points z_1 and z_2 are close enough to Λ_k to be hyperbolic periodic points. Furthermore, if d is sufficiently small the associated stable and unstable manifolds for each point meet transversely (see Figure 27).

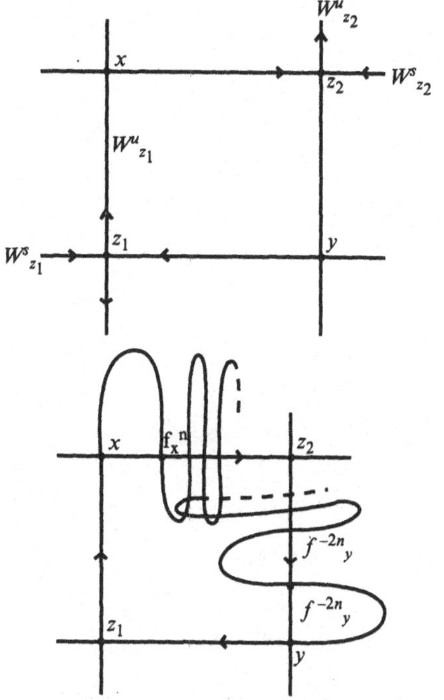

Figure 27: The proof of Proposition 6.3

Thus z_1, z_2 are actually (transverse) homoclinic points. □

Notes

All of the results on the existence of periodic points and homoclinic points come from Katok's 1980 paper [Katok]. The construction of generalized Smale horse-shoes from homoclinic points is originally due to Smale. The continuity properties of the entropy were given by Yomdin in his 1987 article [Yomdin] (and concisely summarized in Gromov's Bourbaki seminar [Gromov]). The examples of discontinuities in the various entropies are variants on examples of Misiurewicz and Mañé.

Chapter 7

Stable manifolds and more measure theory

In this text I have chosen to concentrate on results with a topological flavor. However, in this last chapter, we will attempt to describe part of Pesin's important work on *absolute continuity.* Since the details of the proofs are far beyond the scope of this book, we shall simply concentrate on the statements, and refer the reader to [Kat-Str] or [Pug-Shu] for more complete accounts.

7.1 Stable and unstable manifolds.

In the previous chapter we described the stable and unstable manifolds associated to a periodic point. In this section we shall show how this definition can be extended so that stable and unstable manifolds can be defined for any point $x \in \Lambda$ in a Pesin set.

Definition. Given a non-empty Pesin set $\Lambda = \Lambda(\lambda, \mu, \epsilon)$ (with $\lambda, \mu \gg \epsilon > 0$) we shall define the (local) *stable (unstable) manifolds* through any point $x \in \Lambda$ by

$$W^u_\delta(x) = \{y \in M \mid d(f^{-n}x, f^{-n}y) \leq \delta e^{-(\mu-\epsilon)n},\ n \geq 0\}$$

$$(W^s_\delta(x) = \{y \in M \mid d(f^n x, f^n x) \leq \delta e^{-(\lambda-\epsilon)n},\ n \geq 0\}),$$

for some small $\delta > 0$.

The main result on these stable and unstable manifolds is the following.

Proposition 7.1 (Stable manifold theorem) . Let $f: M \longrightarrow M$ be $C^{1+\alpha}$ and let $\Lambda = \Lambda(\lambda, \mu, \epsilon)$ be a Pesin set (with $\lambda, \mu \gg \epsilon > 0$). There exists $\epsilon_0 > 0$ such that for $x \in \Lambda_k$ ($k \geq 1$) and $\delta = \epsilon_0 e^{-\epsilon k}$:

(a) $W^s_\delta(x)$, $W^u_\delta(x)$ are C^1 submanifolds of M;

(b) $T_x W^s_\delta(x) = E^s_x$, $T_x W^u_\delta(x) = E^u_x$.

(In fact, the stable and unstable manifolds $W^s_\delta(x)$, $W^u_\delta(x)$ are diffeomorphic to discs D^ℓ, D^n whose dimensions are given by $\ell=\dim E^s$, $n=\dim E^u$, as illustrated in Figure 28.)

We shall give a sketch of the proof of Proposition 7.1 in section 7.5.

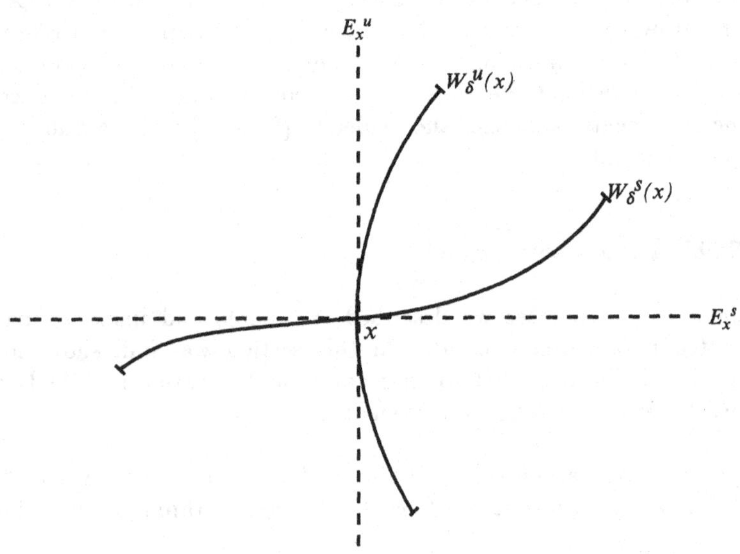

Figure 28: Stable and unstable manifolds

Remark. We can extend these (local) stable and unstable manifolds to global stable and unstable manifolds by replacing $W^s_\delta(x)$ by $\bigcup_{n\geq 0} f^{-n} W^s_\delta(f^n x) = W^s(x)$, and replacing $W^u_\delta(x)$ by $W^u(x) = \bigcup_{n\geq 0} f^n W^u_\delta(f^{-n} x)$.

We now give a few examples to clarify this result.

Examples. (i) Let $M=\mathbb{R}^2/\mathbb{Z}^2$ and let $f: M\longrightarrow M$ be the diffeomorphism defined by $f(x_1, x_2)+\mathbb{Z}^2 = (2x_1+x_2, x_1+x_2)+\mathbb{Z}^2$. As we have noted before, $M = \Lambda_1=\Lambda_2=\cdots=\Lambda$ (with appropriate choices of Liapunov exponents). In this example $\epsilon_0>0$ can be chosen to be any value. The stable and unstable manifolds $W^s_\delta(x)$, $W^u_\delta(x)$ are simply line segments on the torus M which are parallel to eigenvectors of the matrix $\begin{bmatrix} 2 & 1 \\ 1 & 1 \end{bmatrix}$. This is illustrated in Figure 29.

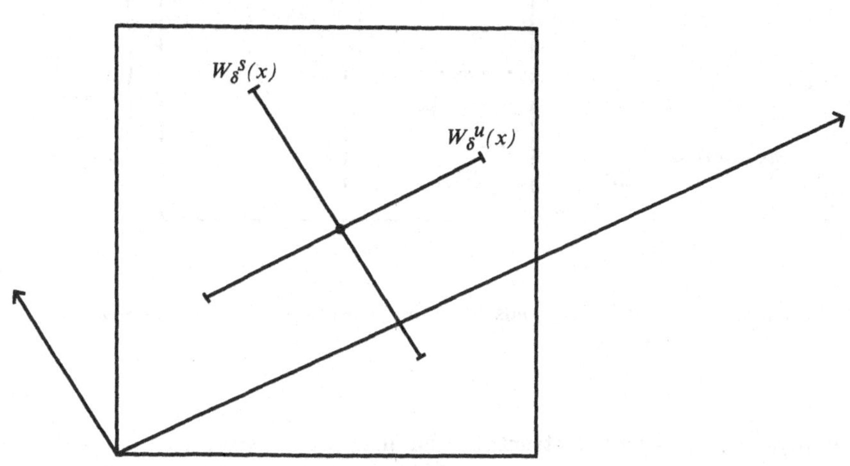

Figure 29: Stable manifolds on the torus

(ii) (Non-uniformly hyperbolic horse-shoes.) If we consider the non-uniformly hyperbolic variations on the standard Smale horse-shoe then the Pesin set will take the form

$$\Lambda \equiv \prod_{-\infty}^{+\infty} \{0,1\} - (\cdots, 0, 0, 0, \cdots)$$

where $\prod_{-\infty}^{+\infty} \{0,1\}$ denotes the usual Cantor set associated with a horse-shoe, and $(\cdots, 0, 0, 0, \cdots)$ denotes the fixed point where we allowed the expansion and contraction to approach unity. The size of the stable and unstable manifolds of points in Λ can be arbitrarily small (corresponding to $x \in \Lambda_k - \Lambda_{k-1}$, for large values of k).

This is illustrated in Figure 30.

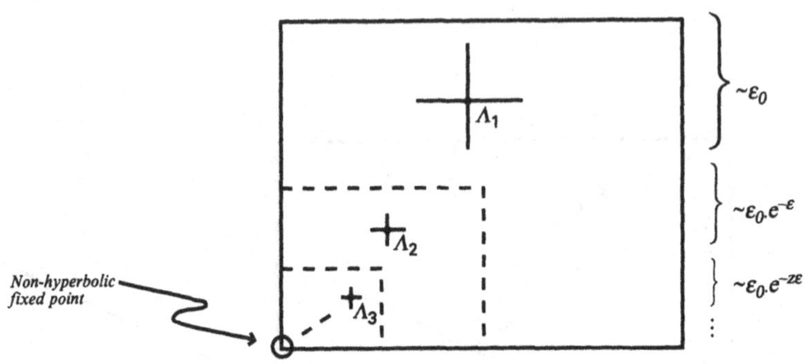

Figure 30: Stable manifolds for non-uniformly hyperbolic horse-shoes

Remark. (Local product structure and maximal measures). By results of Newhouse, every C^∞ map $f: M \longrightarrow M$ has at least one measure of maximal entropy [Newhouse$_1$]. By comparing with the situation for Axiom A diffeomorphisms, the following natural question arises.

An open problem. If $f: M \longrightarrow M$ is a C^∞ surface diffeomorphism with $h_{top}(f)>0$ are there only finitely many ergodic measures of maximal entropy?

We say that a measure of maximal entropy μ has a local product structure if there exist:

(a) a measurable partition $\{R_i\}$ of M;

(b) co-ordinates $R_i = R_i^+ \times R_i^-$ such that for almost all $x = (x_1, x_2)$ the sets $R_i^+ \times \{x_2\}$ and $\{x_1\} \times R_i^-$ coincide with stable and unstable manifolds through x (restricted to R_i);

(c) measures $\mu_i = \mu_i^+ \times \mu_i^-$ on each R_i such that $\mu = \bigcup_i \mu_i$.

Another open problem. If $f: M \longrightarrow M$ is a C^∞ surface diffeomorphism with $h_{top}(f) > 0$ and μ is a measure of maximal entropy then does μ have local product structure?

In the case of Axiom A diffeomorphisms, their are only finitely many ergodic measures of maximal entropy and each has local product structure.

7.2 Equality in the Pesin-Ruelle inequality.

In Chapter 3 we introduced the Pesin-Ruelle inequality for an ergodic measure $\mu \in \mathcal{M}_{erg}$ which related the entropy $h_{meas}(\mu)$ of this measure to its Liapunov exponents λ_i by

$$h_{meas}(\mu) \le \sum_{\lambda_i > 0} \lambda_i.$$

Historically, this arose as a generalization (observed by Ruelle) of the following more specific result (due to Pesin).

Proposition 7.2 (Pesin) If $f: M \longrightarrow M$ is a $C^{1+\alpha}$ diffeomorphism of a compact manifold and $\mu \in \mathcal{M}_{erg}$ is a *smooth* measure (i.e. absolutely continuous with respect to volume) then

$$h_{meas}(\mu) = \sum_{\lambda_i > 0} \lambda_i.$$

Example. Let $M = \mathbb{R}^2/\mathbb{Z}^2$ and let $f: M \longrightarrow M$ be the diffeomorphism defined by $f(x_1, x_2) + \mathbb{Z}^2 = (2x_1 + x_2, x_1 + x_2) + \mathbb{Z}^2$. Let μ denote the ergodic Lebesgue-Haar measure. We have already previously calculated $h_{meas}(\mu) = \log((3+\sqrt{5})/2)$, $\lambda_1, \lambda_2 = \log((3 \pm \sqrt{5})/2)$, thus we have an equality in the above identity, in keeping with Pesin's result.

This prompts the following natural question:

Problem. What are necessary and sufficient conditions for equality?

The following solution was conjectured by Ruelle, and eventually proved by Ledrappier and Young [Led-You].

Proposition 7.3 If $f: M \longrightarrow M$ is a C^2 diffeomorphism of a compact manifold and $\mu \in \mathcal{M}_{erg}$ then there exists an equality

$$h_{meas}(\mu) = \sum_{\lambda_i > 0} \lambda_i$$

if and only if the measure μ induces a smooth measure on unstable manifolds $W^u(x)$.

Example (The solenoid). Let $T \subseteq S^3$ represent a solid torus in the 3-sphere and let $f: S^3 \longrightarrow S^3$ be a diffeomorphism which when restricted to $f: T \longrightarrow T$ 'doubles' the torus, as illustrated in Figure 31.

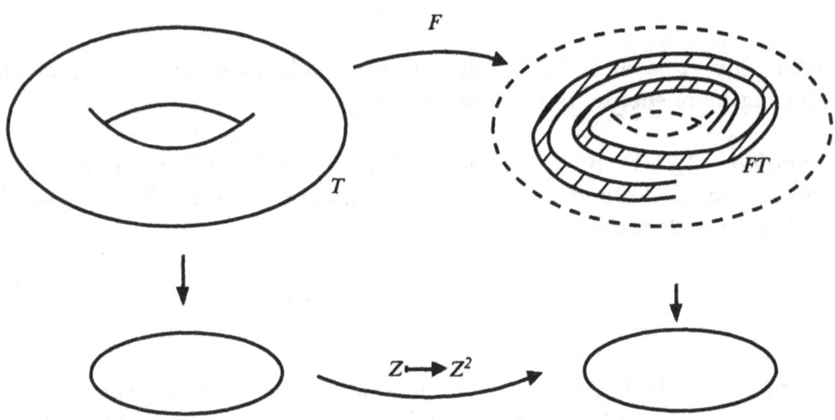

Figure 31: A solenoid

Let $\Lambda = \bigcap_{n=0}^{+\infty} f^n T$ denote the recurrent part of T then $f: \Lambda \longrightarrow \Lambda$ projects to a doubling map on the unit circle S^1 and is locally homeomorphic to a product of the Cantor set $\prod_0^{+\infty} \{0,1\}$ and a one-dimensional manifold. The unstable manifolds for f are contained within Λ. There is a unique ergodic measure μ supported on Λ which locally takes the form $\mu = \left(\frac{1}{2}, \frac{1}{2}\right)^{\mathbb{Z}^+} \times$(Lebesgue measure).

The entropy of $f: \Lambda \longrightarrow \Lambda$ with respect to the measure μ must be the same as the entropy of the doubling map on S^1 with respect to Lebesgue measure. It is easy to check that this is precisely $h_{\text{meas}}(\mu) = \log 2$. The diffeomorphism has three Liapunov exponents. One of these is $\log 2 > 0$ (which corresponds locally to the expanding direction along the circle) and the other two are negative (corresponding to the contraction needed to map $f(T) \subset T$).

7.3 Foliations and absolute continuity.

One of the original uses of the stable and unstable manifolds was to study the ergodic properties of smooth measures. The easiest way to establish ergodicity is to use the following converse to the Birkhoff ergodic theorem.

Lemma 7.1 (Converse Birkhoff). For any f-invariant measure $\mu \in \mathcal{M}_{\text{inv}}$ the limits

$$\begin{cases} C_F^+(x) = \lim_{N \to +\infty} \frac{1}{n} \sum_{n=0}^{N-1} F(f^n x) \\ C_F^-(x) = \lim_{N \to +\infty} \frac{1}{n} \sum_{n=0}^{N-1} F(f^{-n} x) \end{cases}$$

a.a.(m) $x \in M$, exist and are equal for all $F \in C^0(M)$. Furthermore, the f-invariant measure μ is *ergodic* (i.e. $\mu \in \mathcal{M}_{\text{erg}}$) if and only if $C_F^+(x)$ and $C_F^-(x)$ are independent of x.

This result is an easy consequence of the Birkhoff ergodic theorem (Theorem 1.2). For a proof, we refer the reader to [Hopf], or [Sinai$_2$]. Of course, by the Birkhoff ergodic theorem (Theorem 1.2) we know that $C_F^+(x)$ $(= C_F^-(x)\ ) \equiv \int F dm$, for a.a.$(m)$ $x \in M$, when $m \in \mathcal{M}_{\text{erg}}$.

We now want describe a standard method to prove ergodicity of certain measures $m \in \mathcal{M}_{\text{inv}}$ using stable and unstable manifolds and Lemma 7.1. Assume that we take two points $x_1, x_2 \in M$ such that the associated stable manifolds $W_\delta^s(x_1)$ and $W_\delta^s(x_2)$ intersect some common unstable manifold $W_\delta^u(z)$ through a point $z \in \Lambda$. We denote these points of intersection by y_1, y_2, respectively. This is illustrated in Figure 32.

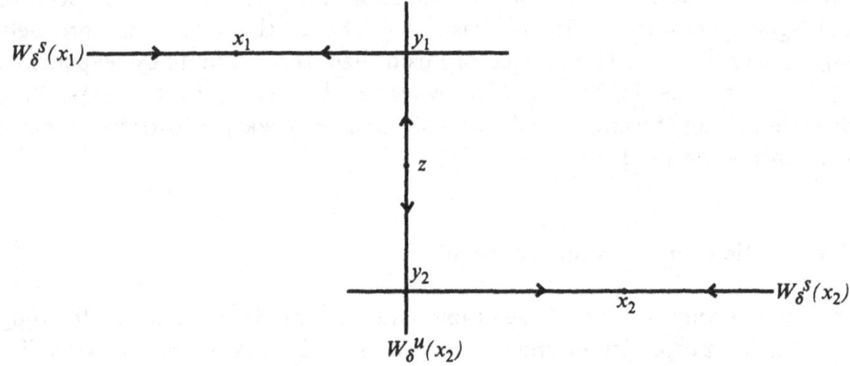

Figure 32: Ergodicity via stable manifolds

Assume that the limits $C_F^+(y_1)=C_F^-(y_1)$ and $C_F^+(y_2)=C_F^-(y_2)$ are both well-defined. We can then observe the following.

(a) Since $y_1,y_2 \in W_\delta^u(z)$ they must have the same limits $C_F^-(y_1)=C_F^-(y_2)$. Furthermore, since $d(f^{-n}y_1,f^{-n}y_2) \longrightarrow 0$ as $n \longrightarrow +\infty$ the limits must also converge, i.e. $C_F^+(y_1) = C_F^-(y_1) = C_F^-(y_2) = C_F^+(y_2)$.

(b) For each of the points $y_i \in W_\delta^s(x_i)$ $(i=1,2)$ a similar reasoning gives that $C_F^+(y_i)=C_F^+(x_i)$ since $d(f^n x_i,f^n y_i) \longrightarrow 0$ as $n \longrightarrow +\infty$ $(i=1,2)$. Therefore, we can conclude that $C_F^+(x_1) = C_F^+(x_2)$.

Using Lemma 7.1, the above argument can be used to prove ergodicity if: *(i) there exists of a 'network' of such stable and unstable manifolds linking almost all (with respect to* m*)* points; and *(ii) that no measure is 'lost' by translating to points along stable and unstable manifolds.*

<u>*Note*</u>. This is usually called the Hopf-argument, after Hopf's use of a similar argument to prove the ergodicity of geodesic flows [Hopf]. It has also been very successfully applied to Billiard models (see [Kat-Str]).

The need to formulate condition (ii) more rigorously introduces the notion of absolute continuity.

Definition. We say that a family of unstable manifolds is ***absolutely continuous*** if whenever we choose sufficiently small discs D_1, D_2 transverse to $\{W^u(x) \mid x \in \Lambda\}$ and whenever the map $y_1 \longrightarrow y_2$ with $y_1 = D_1 \cap W^u(x)$, $y_2 = D_2 \cap W^u(x)$, is well defined it has the property that sets of zero volume on D_1 are mapped to sets of zero volume on D_2.

This is illustrated in Figure 3.3. There is a similar definition for a family of stable manifolds $W^s(x)$.

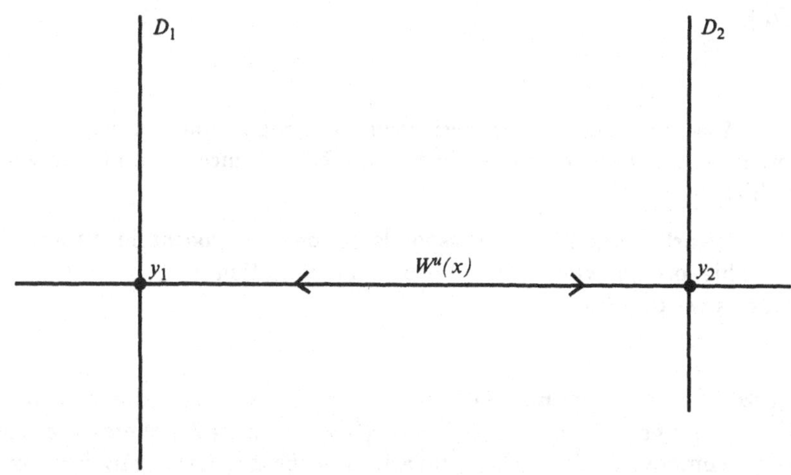

Figure 33: Absolute continuity

Pesin showed that these families are always absolutely continuous:

Theorem 7.1 (Pesin). For any $C^{1+\alpha}$ diffeomorphism of a compact manifold the families of stable and unstable manifolds are absolutely continuous (on each Pesin set).

For the proof, we refer the reader to [Pug-Shu] or [$\text{Pesin}_{1,2}$].

7.4 Ergodic components.

We now come to the important application of absolute continuity to the understanding of the ergodic structure of smooth invariant measures. The basic ideas are contained in the following special case.

Proposition 7.4 Let $f: M \longrightarrow M$ be a $C^{1+\alpha}$ diffeomorphism of a compact surface. Assume that there exists an absolutely continuous measure $m \in \mathcal{M}_{\text{inv}}$ which satisfies $h_{\text{meas}}(m) > 0$. Then

(a) M can be written as a *countable* union $M = \bigcup_{n=0}^{+\infty} \Omega_n$ (up to a set of zero measure, with repect to m) ;

(b) $f(\Omega_n) = \Omega_n$ and $f: \Omega_n \longrightarrow \Omega_n$ is ergodic, with respect to m (normalized on Ω_n).

We shall sketch the derivation of Proposition 7.4 from Pesin's theorem on absolute continuity in Section 7.6. A nice account appears in [Pug-Shu].

Clearly, ergodicity corresponds to *one component* in Proposition 7.4. This occurs when the 'network' in condition (ii), of section 7.4, includes a set of full measure.

Remarks. (i) For the manifold $M = \mathbb{R}^2/\mathbb{Z}^2$ and the diffeomorphism $f: M \longrightarrow M$ defined by $f(x_1,x_2)+\mathbb{Z}^2 = (2x_1+x_2, x_1+x_2)+\mathbb{Z}^2$ there is a *single* ergodic component (i.e. the f-invariant Lebesgue-Haar measure m is actually ergodic.). In this case any two points can be linked by one stable and one unstable manifold (in the sense of condition (i)).

(ii) Without the associated stable and unstable manifolds there may be *uncountably* many ergodic components. Consider the simple example of the diffeomorphism $f: \mathbb{R}^2/\mathbb{Z}^2 \longrightarrow \mathbb{R}^2/\mathbb{Z}^2$ defined by $f(x_1,x_2)+\mathbb{Z}^2 = (x_1, x_1+x_2)+\mathbb{Z}^2$. This preserves Lebesgue-Haar measure m, but is not ergodic. In fact, the torus decomposes into an uncountable family of invariant circles (parametrized by the first co-ordinate x_1) which are rotated through an angle x_1. Thus, restricted to these circles the diffeomorphism preserves the associated (one-dimensional) Lebesgue measure and is ergodic whenever x_2 is irrational.

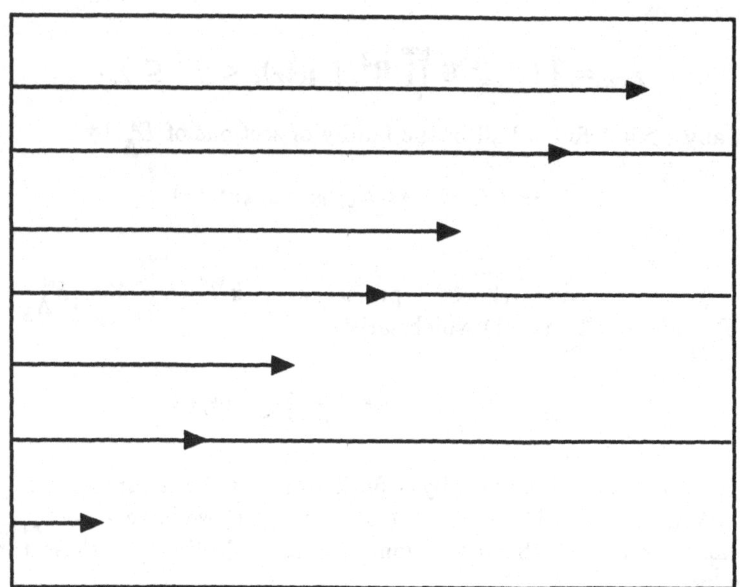

Figure 34: Infinitely many ergodic components

7.5 Proof of Stable manifold theorem.

For the construction of the stable manifolds we shall follow **[Mañé$_1$]**. In particular, we shall use the Perron (implicit function) approach rather than the alternative Hadamard (Graph transform) method (cf.**[Anosov]**, p.23).

For convenience assume Λ is trivialized (using charts) such that $\Lambda \subseteq \overline{\Lambda} \subseteq U \subseteq \mathbb{R}^d$. For later use, choose $\lambda, \mu \gg \epsilon > 0$ with

(i) $(\mu + 3\epsilon) < \lambda$;

(ii) $\epsilon < \frac{\lambda\alpha}{3}$;

Step 1. Fix a constant $0 < \gamma < d(\overline{\Lambda}, \mathbb{R}^d - U)$. We can introduce a Banach space of sequences (of vectors) as follows:

$$\Sigma = \{(\xi_r)_{r=0}^{+\infty} \in \prod_0^{+\infty} \mathbb{R}^d \mid \|(\xi_r)\| < +\infty\}$$

where $\|(\xi_r)\| = \sup_r [\, |\xi_r| . e^{(\lambda-\epsilon)r}]$ and we denote the ball about the origin of radius γ by

$$\Sigma_\gamma = \{ (\xi)_{r=0}^{+\infty} \in \prod_0^{+\infty} \mathbb{R}^d \mid \|(\xi_r)\| < \gamma \} \subseteq \Sigma.$$

Given any $\delta>0$ define a ball in the family of sections of E^s_Λ by

$$\Gamma_\delta = \{ (x,v) \mid x \in \Lambda_k,\ v \in E^s_x,\ \|v\| < \delta \}.$$

Step 2. We can choose projections $\pi^{u,k}: T_{\Lambda_k} M \longrightarrow E^u_{\Lambda_k}$ and $\pi^{u,k}: T_{\Lambda_k} M \longrightarrow E^u_{\Lambda_k}$ $(k \geq 1)$ which satisfy

$$\left\|\pi^{s,k+n}_{f^n x}\right\|, \left\|\pi^{u,k+n}_{f^n x}\right\| \leq e^{\epsilon k} e^{\epsilon n}$$

(perhaps up to multiplication by a fixed constant, by assumption (c) on a Pesin set in Section 4.1). If $x \in \Lambda_k$ (for some $k \geq 1$) we have $f^n x \in \Lambda_{k+|n|}$, for $n \geq 0$, and we denote the restrictions of the projections to these tangent spaces by

$$\begin{cases} \pi^{s,k+|n|}_{f^n x}: T_{f^n x} M \longrightarrow E^s_{f^n x} \\ \pi^{u,k+|n|}_{f^n x}: T_{f^n x} M \longrightarrow E^u_{f^n x} \end{cases}$$

For any $\delta>0$, define a map $F: \Gamma_\gamma \times \Sigma_\delta \longrightarrow \Sigma$ as follows

$$\begin{cases} F((x,v), (\xi_m))_0 = v - \sum_{j=0}^{+\infty} (D_x f^{(j+1)})^{-1} \pi^{u,k+(j+1)}_{f^{j+1} x} \rho\, (f^j x,\ \xi_j) \\ F((x,v), (\xi_m))_n = (D_x f^n)(v) + S_1 + S_2 \ (n \geq 1) \end{cases}$$

where we denote:

(a) $S_1 = \sum_{j=0}^{n-1} \left(D_{f^{j+1} x} f^{(n-1)-j} \right) \pi^{s,k+j+1}_{f^{j+1} x}\ \rho\, (f^{j+1} x,\ \xi_j)$;

(b) $S_2 = -\sum_{j=n}^{+\infty} \left(D_{f^n x} f^{j-(n-1)} \right)^{-1} \pi^{u,k+j+1}_{f^{j+1} x} \rho\, (f^j x,\ \xi_j)$; and

(c) $\rho(x,v) = f(x+v) - f(x) - (D_x f)(v)$

where ρ is well-defined by the choice of γ sufficiently small.

Step 3. We still need to show that $F(\Gamma_\gamma \times \sum_\delta) \subseteq \sum$. For fixed $\gamma > 0$, we can bound the above expressions S_1 and S_2 , as follows.

(a) $|S_1|$

$$\leq (Const.)\, e^{2\epsilon k} \sum_{j=0}^{n-1} [e^{\epsilon(j+1)} e^{-\lambda(n-1-j)}][e^{\epsilon(j+1)}][\,\|(\xi_r)\| e^{-\lambda j}]^{(1+\alpha)}$$

$$\leq (Const.)\, e^{2\epsilon k} \|(\xi_r)\|^{(1+\alpha)} \left(\frac{e^{-(\lambda-\epsilon)n} e^{\epsilon+\lambda}}{1 - e^{-\lambda\alpha+\epsilon}} \right)$$

$$= O(e^{-(\lambda-\epsilon)n}), \text{ and}$$

(b) $|S_2|$

$$\leq (Const.)\, e^{2\epsilon k} \sum_{j=n}^{+\infty} [e^{\epsilon(j+1)}\, e^{-\mu(j-n+1)}][\, e^{\epsilon(j+1)}][\|(\xi_r)\| e^{-\lambda j}]^{(1+\alpha)}$$

$$\leq (Const.)\, e^{2\epsilon k} \|(\xi_r)\|^{(1+\alpha)} \frac{e^{n(2\epsilon-\lambda(1+\alpha))} e^{-(\mu-2\epsilon)n}}{1 - e^{-(\mu-2\epsilon+(1+\alpha)\lambda)}}$$

$$= O(e^{-(\lambda-\epsilon)n})$$

(Where *Const.* represents a constant value whose precise value we do not need to keep track of.)

In particular, these bounds tell us that $F((x,v),(\xi_r)) \in \sum$, as claimed in Step 2, and that the map F is continuous. Furthermore, since we have assumed that $f: M \longrightarrow M$ is $C^{1+\alpha}$ we have

$$\|\rho(x, v+\omega) - \rho(x,v)\| \leq (Const.')\, \|\omega\|^{1+\alpha}.$$

Returning to the definition of F: $\Gamma_\gamma \times \sum_\delta \longrightarrow \sum$ we can deduce

$$|F((x,v), (\xi_r))_n - F((x,v), (\xi_r + \xi_r'))_n| e^{\lambda n} \leq C\|(\xi_r')\| e^{2\epsilon k}$$

where $C = C(\lambda,\epsilon,\alpha)$ is independent of k. Therefore, for fixed $(x,v) \in \Gamma_\gamma$ we have, that

$$F((x,v),\, .\,): \textstyle\sum_\delta \longrightarrow \sum \text{ is } C^{1+\alpha}, \qquad (7.1)$$

with derivative of order $O(e^{2\epsilon k})$.

Step 4 (implicit function theorem). We need the following lemma.

Lemma 7.3 For $x\in\Lambda_k$ and $v\in E^s_x$ with $\|v\|<\delta<\rho$ the following are equivalent for $y\in\mathbb{R}^d$:

(a) $\pi_x^{s,k}(y) = x+v$, and $\|f^n y - f^n x\| \leq \delta\, e^{-(\lambda-\epsilon)n}$, $\forall n\geq 0$;

(b) $\exists(\xi_r)_{r=0}^{+\infty}\in\sum_\delta$ such that $\xi_0=y-x$ and $F((x,v),(\xi_r))=(\xi_r)$.

The proof is not very difficult, and we refer the reader to [Mañé$_1$], p.574, for details. We observe that the existence of stable manifolds is equivalent to (a). Therefore to complete the proof of the lemma it suffices to prove (b) (which we shall do below). The proof of (b) only requires the implicit function theorem and previous estimates.

Define a map Φ: $\Gamma_\gamma\times\sum_\delta \longrightarrow \sum$ by

$$\Phi\ ((x,v),\ (\xi_r)) = (\xi_r) - F((x,v)\ (\xi_r)),$$

then

(i) $\Phi((x,0),0) = 0$, and

(ii) $D_3\Phi((x,0),0) = I - D_3\ F((x,0),0)$.

By the estimate (7.1) in Step 3 we have that whenever $\|(\xi_r)\| \leq (Const.)e^{-2\epsilon k}$ then $\|D_3 F((x,v),0)\| < 1$. In particular, we have that $\|D_3\Phi((x,v),0)-I\| <1$ provided $\delta \leq (Const.)\ e^{-2\epsilon k}$. We can apply the implicit function theorem to F to see that part (b) of Lemma 7.3 is satisfied (and thus part (a) of Lemma 7.3).

To realize the the stable manifolds, we first define

$$V^s_\delta(x) = \{\ y\in\mathbb{R}^d \mid \|f^n x - f^n y\| \leq \delta\, e^{-(\lambda-\epsilon)n},\ n\geq 0\}$$

and then using the exponential map (see Appendix B) we define $W^s_\delta(x) = \exp_x V^s_\delta(x)$.

We can similarly construct $V^u_\delta(x)$ and the unstable manifold $W^u_\delta(x) = \exp_x V^u_\delta(x)$. This completes the proof of Proposition 7.1. □

7.6 Ergodic components and absolute continuity.

We shall briefly outline the key idea in the proof of the Proposition 7.4.

Since we have made the assumption $h_{\text{meas}}(m)>0$ we have the existence of a Pesin set Λ with $m(\Lambda)=1$ (by Proposition 4.2). By Pesin's theorem (Theorem 7.1) the stable manifolds are absolutely continuous, which implies that we can choose $x \in \Lambda$ with

$$m\Big(\bigcup_{y \in W^u_\delta(x)} W^s_\delta(y) \cap \Lambda \Big) > 0.$$

(for example, it happens whenever we choose $x \in \Lambda$ to be a density point of the measure). We can define an equivalence relation on Λ by:

$$x \sim y \text{ if } \begin{cases} \exists\ x_0, \cdots, x_n \in \Lambda \text{ with } x_0 = x,\ x_n = y \textit{ and} \\ W^u_\delta(x_i) \cap W^s_\delta(x_{i+1}) \neq \phi \end{cases}$$

To make this definition completely rigorous, we would need more technical details, which can be found in [Pug-Shu]. We let $[x]$ denote the associated equivalence class of $x \in \Lambda$. By construction, we have the estimate

$$m([x]) \geq m\Big(\bigcup_{y \in W^u_\delta(x)} W^s_\delta(y) \cap \Lambda \Big) > 0.$$

In particular, there can be at most a countable number of such equivalence classes since they are disjoint and have non-zero measure. We define Ω_n, $n \geq 1$, to be these equivalence classes.

It is easy to see from the construction that $f(\Omega_n)=\Omega_n$. The ergodicity of the restriction $f\colon \Omega_n \longrightarrow \Omega_n$ to each component Ω_n follows by the Hopf arguments explained in Section 7.3. □

Notes

The account of stable manifold theory we give is adapted from a 1983 article of Mañé [Mañé$_1$].

The necessary and sufficient condition for equality in the Pesin-Ruelle inequality was shown by Ledrappier and Young in their 1985 article [Led-You].

The use of the inverse Birkhoff theorem to prove ergodicity was used in Hopf's 1939 book to prove ergodicity of geodesic flows (and it is the basis of Sinai's approach to proving ergodicity for dispersive Billiards). Absolute continuity was originally proved by Pesin in [$\text{Pesin}_{1,2}$] (although a more detailed account, with some corrections, appears in the book of Katok-Strelycn). The use of absolute continuity to give only countably many ergodic components is due to Pesin, and a nice account is given in the article of Pugh-Shub [Pug-Shu].

Appendix A

Some preliminary measure theory

This appendix to the main notes is intended to explain some of the conventions in measure theory which we have taken for granted in the notes. An admirable reference for this material is the book [Parthasarathy].

Fortunately, in practice this abstract formalism usually doesn't appear in theorems and their proofs, where the underlying formalism is taken for granted and more workable conventions are adopted. (This is perhaps much the same philosophy as when $\mathbb{R}$ is constructed as equivalence classes of Cauchy sequences of rationals.)

A.1 Algebras and σ-algebras.

Assume we are given an abstract set X.

Definition. An ***algebra*** $\mathcal{O}$ for X is collection of subsets of X such that:

(i) $\emptyset, X \in \mathcal{O}$;

(ii) $A_1, \cdots, A_n \in \mathcal{O} \Rightarrow \bigcup_{i=1}^{n} A_i \in \mathcal{O}$; and

(iii) $A_1, \cdots, A_n \in \mathcal{O} \Rightarrow \bigcap_{i=1}^{n} A_i \in \mathcal{O}$.

Remark. An algebra $\mathcal{O}$ is *usually* assumed closed under a finite number of set theoretic operations; intersection $\cap$, union $\cup$, complement ^{c}, symmetric difference Δ, etc. Thus the above definition is *weaker* than the standard definition, but we prefer this version so that the following makes sense.

Most important example. The most important example for our purposes is where X is a topological space and $\mathcal{O}$ is the topology.

<u>Definition</u>. A *σ-algebra* $\mathfrak{B}$ for X is a collection of subsets of X such that

(i) ϕ, $x \in \mathfrak{B}$;

(ii) $B_1, B_2, \cdots \in \mathfrak{B} \Rightarrow \bigcup_{i=1}^{+\infty} B_i \in \mathfrak{B}$; *and*

(iii) $B \in \mathfrak{B} \Rightarrow B^C \in \mathfrak{B}$.

In particular, σ-algebras are also algebras since we can write $\bigcap_i B_i = (\bigcup_i B_i^C)^C \in \mathfrak{B}$.

Trivial examples of σ-algebras occur for any set X and the choices $\mathfrak{B} = \{\phi, X\}$, $\mathfrak{B} = \{B \subseteq X\}$.

Given an algebra $\mathfrak{O}$ we can generate an associated σ-algebra $\mathfrak{B} = \mathfrak{B}(\mathfrak{O})$ by simply adding in all countable unions and complements of existing sets (and also of those we generate). When the algebra $\mathfrak{O}$ that we start from is a topology on X then $\mathfrak{B} = \mathfrak{B}(\mathfrak{O})$ is called the *Borel σ-algebra.*

<u>Remark</u>. In general we expect that our σ-algebra does not exhaust all subsets of X, i.e. there exist subsets $Y \subseteq X$ such that $Y \notin \mathfrak{B}$. For example, if $X = [0,1]$, and $\mathfrak{B}$ is the Borel σ-algebra then there exists $Y \notin \mathfrak{B}$.

<u>Definition</u>. A pair $(X,\mathfrak{B})$ is called a *measurable space.*

A.2. Measures.

Assume we are given a measurable space $(X,\mathfrak{B})$.

<u>Definition</u>. A *measure* m on $(X,\mathfrak{B})$ is a map m: $\mathfrak{B} \longrightarrow \mathbb{R}^+$ such that

(i) $m(\emptyset) = 0$; and

(ii) $B_1, B_2, \cdots \in \mathfrak{B}$ $(B_i \cap B_j = \phi,\ i \neq j) \Rightarrow m(\bigcup_{i=1}^{+\infty} B_i) = \sum_{i=1}^{+\infty} m(B_i)$

In addition, m is called a *probability measure* if $m(X) = 1$.

<u>Examples</u>. (i) Consider the case where $X = M$ is a Riemannian manifold and m is a normalized 'volume' (e.g. $X = [0,1]$ and m is Lebesgue measure; or $X = \mathbb{R}^n/\mathbb{Z}^n$ and m is a Lebesgue-Haar measure).

(ii) (Dirac measure). If we fix any point $x \in X$ we can define a probability measure $m: \mathcal{B} \longrightarrow \mathbb{R}^+$ by

$$m(B)=\begin{cases}1 & \text{if } x\in B\\ 0 & \text{if } x\notin B\end{cases}$$

The usual convention is to write $M=\delta_x$.

A.3 Presentation of measures.

Rather than specifying the measure on all sets in the σ-algebra $\mathcal{B}$ it is usually enough to specify it on somewhat smaller families. The basic result in this direction is the following.

Theorem (Extension theorem). Assume $\mathcal{O}$ is an algebra and $\mathcal{B} = \mathcal{B}(\mathcal{O})$ the σ-algebra it generates. Then any map $m: \mathcal{O} \longrightarrow \mathbb{R}^+$ which satisfies:

(i) $m(\emptyset)=0$;

(ii) $A_1, \cdots, A_n \in \mathcal{O}$ disjoint $\Rightarrow m(\bigcup_{i=1}^{n} A_i) = \sum_{i=1}^{n} m(A_i)$,

extends (uniquely) to a measure on X.

Familiar example. Let $X = [0,1]$ and $\mathcal{S} = \{[a,b) \mid 0 \leq a < b \leq 1\}$ then specifying $m([a,b)) = (b-a)$ gives a Lebesgue measure on the Borel σ-algebra by the above theorem.

A.4 Measurability.

Since only subsets of X in $\mathcal{B}$ are in some sense 'acceptable' we have to place restrictions on maps from X.

Definitions. If $\mathcal{R}$ is the Borel σ-algebra for $\mathbb{R}$ then a map $f: X \longrightarrow \mathbb{R}$ is called *measurable* if $f^{-1}(\mathcal{R}) \subseteq \mathcal{B}$. A transformation $T: X \longrightarrow X$ is called *measurable* if $f^{-1}(\mathcal{B}) \subseteq \mathcal{B}$.

Main example. If X is a topological space and $\mathcal{B}$ is the Borel σ-algebra then continuous maps $f: X \longrightarrow \mathbb{R}$ and $T: X \longrightarrow X$ are automatically measurable.

Notation. Let $\mathcal{L}^0(X,\mathcal{B}) = \{f: X \longrightarrow \mathbb{R} \mid f \text{ is measurable}\}$ be the space of measurable functions on X. Let $\mathcal{L}^1(X,\mathcal{B},m) = \{f: X \longrightarrow \mathbb{R} \mid \int |f| \, dm < +\infty,$

f is measurable} be the space of integrable functions on X. Let $\mathcal{L}^\infty(X, \mathcal{B}) = \{f\colon X \longrightarrow \mathbb{R} \mid f \textit{ is measurable and bounded a.e.}(m)\}$ be the space of functions bounded almost everywhere.

Note. To make $\mathcal{L}^1$ into a Banach space we need to identify functions equivalent under the equivalence relation: $f \sim g$ if f and g differ only on a set of zero measure. We can write $\mathcal{L}^0/\sim \; = L^0$, $\mathcal{L}^1/\sim \; = L^1$, and $\mathcal{L}^\infty/\sim = L^\infty$

A.5 Comparing measures.

In general we shall assume that the measurable space $(X,\mathcal{B})$ is given, and consider the set $\mathcal{M} = \{m \mid m \textit{ is probability measure}\}$.

Definition. Let m_1, $m_2 \in \mathcal{M}$ be two probability measures.

(i) We say that m_1 is *absolutely continuous* with respect to m_2 (denoted $m_1 \ll m_2$) if $\forall\ B \in \mathcal{B}$, $m_2(B) = 0 \Rightarrow m_1(B) = 0$.

(ii) We say that m_1, m_2 are *equivalent* (denoted $m_1 \sim m_2$) if $m_1 \ll m_2$ and $m_2 \ll m_1$.

(iii) We say that m_1, m_2 are *mutually singular* (denoted $m_1 \perp m_2$) if $\exists B \in \mathcal{B}$ such that $m_1(B) = 0$, $m_2(B) = 1$.

Remark. If $m_1 \ll m_2$ then the Radon-Nikodym theorem gives that there exists $f \in L^1(X,\mathcal{B},m_2)$ such that $m_1(B) = \int_B f\, dm_2$, $\forall\ B \in \mathcal{B}$. It is the convention to denote $f = \frac{dm_1}{dm_2}$.

Example. Let $X = [0,1]$ and let $\mathcal{B}$ be the usual Borel σ-algebra. We define the measure m_1 to be the usual Lebesgue measure and the measure m_2 to be δ_0 (i.e. the Dirac measure supported on the single point $x=0$). Then the measures m_1, m_2 are mutually singular. If $B=(0,1]$ then $m_1(B)=1$, $m_2(B)=0$.

A.6 Additional structure for topological spaces.

When X is a compact topological space and $\mathfrak{B}$ is the Borel σ-algebra we have more structure for $\mathcal{M}$. This will be the situation for most of our examples.

Definition. We define the *support* of $m \in \mathcal{M}$ (denoted supp(m)) to be the smallest closed set C with $m(C) = 1$.

Example. Let $X = [0,1]$ and let $\mathfrak{B}$ be the Borel σ-algebra. We define m_1 to be the usual Lebesgue measure and m_2 to be the Dirac measure δ_0. In this case, $\text{supp}(m_1) = [0,1]$ and $\text{supp}(m_2) = \{0\}$.

Definition. We can define a topology, called the ***weak* topology***, on $\mathcal{M}$ by $\mu_n \longrightarrow \mu$ as $n \longrightarrow +\infty \Leftrightarrow \int F\, d\mu_n \longrightarrow \int F\, d\mu$ as $n \longrightarrow +\infty$, $\forall\ F \in C^0(X)$ (where $\mu_n, \mu \in \mathcal{M}$, $n \geq 1$).

Theorem (Alaoglu). $\mathcal{M}$ is compact with the weak* topology.

Note. We also observe that $\mathcal{M}$ is non-empty (always containing, for example, the Dirac measure) and convex, i.e. if $m_1, m_2 \in \mathcal{M}$ then $\alpha m_1 + (1-\alpha) m_2 \in \mathcal{M}$, $0 \leq \alpha \leq 1$ (where $[\alpha m_1 + (1-\alpha) m_2]\ (B) = \alpha\, m_1(B) + (1-\alpha) m_2(B)$, $\forall\ B \in \mathfrak{B}$).

Appendix B

Some preliminary differential geometry theory

This appendix is intended to explain some of the basic ideas in differential geometry which we have taken for granted in the main text. There exist a number of excellent modern introductory texts, from amongst which we might suggest the very readable book of Gallot, Hulin and Lafontaine [Ga-Hu-La].

B.1 Differentiable manifolds and maps.

Let M be a compact metric space. We call M a (d-dimensional) C^k *manifold*, where $k \geq 1$, if there exists an open cover $\{U^\alpha\}$ for M and homeomorphisms $x^\alpha\colon U^\alpha \longrightarrow V^\alpha$ onto open sets $V^\alpha \subseteq \mathbb{R}^d$ such that each composition $x^\alpha \circ (x^\beta)^{-1}$ is a C^k map (on neighborhoods of $\mathbb{R}^d$) whenever it is defined.

If a subset $V \subseteq M$ is also a manifold, then we call it a *submanifold* of M. If M is a C^∞ manifold which has, in addition, a C^∞ group operation then it is called a *Lie group* (for example, the torus $\mathbb{T}^d = \mathbb{R}^d/\mathbb{Z}^d$ with the operation $(x+\mathbb{Z}^d,\ y+\mathbb{Z}^d) \mapsto x+y+\mathbb{Z}^d$).

Definition. The maps $x^\alpha = (x_1^\alpha, \cdots, x_d^\alpha)$ are called *local co-ordinates* (or *charts*) for M.

Convention. For local properties of the manifold M we will restrict to one of the neighborhoods U^α and formulate statements using the local co-ordinates (with the implicit understanding that statements are independent of the specific choice of chart).

We call a continuous map $f\colon N \to M$ between two C^k manifolds a C^k *map* if for any charts $x^\alpha\colon U^\alpha \to V^\alpha$ $(U^\alpha \subset M)$ and $y^\beta : \mathcal{U}^\beta \to \mathcal{V}^\beta$ $(\mathcal{U}^\beta \subset N)$ the map $x^\alpha \circ f \circ (y^\beta)^{-1}$ is a C^k map (on neighborhoods of $\mathbb{R}^n$) whenever it is defined. If the map f has a C^k inverse then we call it a C^k *diffeomorphism*. We let $\mathrm{Diff}^k(M)$ = denote the space of all C^k

diffeomorphisms (and $\mathrm{Diff}^{k+\alpha}(M)$, $\alpha>0$, the space of all C^k diffeomorphisms with α-Hölder continuous kth derivatives)

We recall two basic results about the global geometry of a compact C^k manifold M of dimension d.

Proposition (Triangulation). The manifold M can be sub-divided into a finite number of simplices (C^k diffeomorphic to the standard simplex in $\mathbb{R}^d$) which only meet at their faces.

The proof of this proposition is intuitively obvious: first divide up M into polyhedra using $(d-1)$-dimensional hyperplanes in the local co-ordinates; then divide up the polyhedra into simplices.

Proposition (Whitney embedding). There exists a C^k submanifold $N \subset \mathbb{R}^{(2d+1)}$ and a C^k diffeomorphism $f: M \longrightarrow N$.

B.2 Tangent bundles and vector fields.

Let E be a metric space and let $\pi: E \to M$ be a continuous surjective map.

Definition. We call E a *vector bundle* if:

(i) for each $x \in M$ the *fiber* $E_x = \pi^{-1}(x)$ is a vector space;

(ii) the neighborhoods U^α can be chosen so that there are continuous maps $\sigma^\alpha: U^\alpha \to E$ such that $\pi \circ \sigma^\alpha = id| U^\alpha$.

Given two vector bundles and maps $\pi: E \to M$, $\pi': E' \to M$, over the same manifold M, we can define the *Whitney sum* (or *direct sum*) $E'' = E' \oplus E$ to be the vector bundle (over M, again) whose fibers E''_x, $x \in M$, are the direct sum $E'_x \oplus E_x$ as vector spaces of the corresponding fibers for E' and E.

Our main example of a vector bundle is the tangent bundle TM, for which the fibres (usually denoted $T_x M$, $x \in M$) are the vector spaces of linear maps $\alpha'(0)$: $C^k(M) \longrightarrow \mathbb{R}$ associated to all C^k curves $\alpha: (-\epsilon, \epsilon) \longrightarrow M$ (with $\alpha(0)=x$) by $\alpha'(0): f \mapsto \frac{d}{dt}(f \circ \alpha)|_{t=0}$.

Definition. The *tangent bundle* is the union $TM = \bigcup_{x \in M} T_x M$.

If $V \subset M$ is a submanifold then $T_x V \subset T_x M$, $x \in M$, is a subspace. For subsets $\Lambda \subset M$ we can define $T_\Lambda M = \bigcup_{x \in \Lambda} T_x M$.

Bases. The local co-ordinates $x^\alpha = (x_1^\alpha, \cdots, x_d^\alpha)$ give a basis for $T_{U^\alpha} M$ which is usually denoted by

$$\left(\frac{\partial}{\partial x_i^\alpha}\right): \mathrm{f} \mapsto \frac{\partial (f \circ (x^\alpha)^{-1})}{\partial x_i^\alpha}(0), \text{ where } f \in C^k(M, \mathbb{R}).$$

(In particular, this shows that each fiber $T_x M$, $x \in M$, is d-dimensional.)

Given a C^k map $f: M \to M$ and charts $x^\alpha: U^\alpha \to V^\alpha$ and $x^\beta: U^\beta \to V^\beta$ with $x \in U^\alpha$, $fx \in U^\beta$ the derivative $D_x f: T_x M \to T_{fx} M$ is the linear map represented by the matrix

$$\left(\frac{\partial (x_j^\beta \circ f \circ x_i^\alpha)}{\partial x_j^\alpha}(0)\right)_{i,j=1}^d$$

with repect to the bases $\left\{\frac{\partial}{\partial x_i^\alpha}\right\}, \left\{\frac{\partial}{\partial x_j^\beta}\right\}$.

A *C^k vector field* is a continuous map $X: M \longrightarrow TM$ with $X(x) \in T_x M$ defined by

$$X(x) = \sum_{i=1}^d v_i(x) \frac{\partial}{\partial x_i^\alpha}, \quad v_i \in C^k(M, \mathbb{R})$$

(in local co-ordinates). We let $\mathfrak{X}^k(M)$ denote the space of all C^k vector fields. *Henceforth, we shall only concern ourselves with C^∞ manifolds and C^∞ vector fields.*

Forms and de Rham cohomology. We denote by $\Lambda^1(M)$ the dual space to the linear topological space $\mathfrak{X}^\infty(M)$, and the continuous linear functionals $\omega \in \Lambda^1(M)$ are called the *1-forms.* The space $\Lambda^1(M)$ has a natural dual basis $\{dx_i^\alpha\}_{i=1}^d$, defined by

$$dx_i^\alpha\left(\frac{\partial}{\partial x_j^\alpha}\right) = \begin{cases} 1 & \text{if } i=j \\ 0 & \text{if } i \neq j \end{cases} \quad \text{(in local co-ordinates).}$$

We define the space $\Lambda^k(M)$, $1 \le k \le d$, of *k-forms* to be the vector space spanned by $\rho(x)\, dx_{i_1}^\alpha \wedge \ldots \wedge dx_{i_k}^\alpha$, $\rho \in C^\infty(M, \mathbb{R})$, subject to two basic axioms:

$$dx_{i_1}^\alpha \wedge \ldots \wedge (\rho\, dx_{i_m}^\alpha) \wedge \ldots \wedge dx_{i_k}^\alpha = \rho\, dx_{i_1}^\alpha \wedge \ldots \wedge dx_{i_m}^\alpha \ldots \wedge dx_{i_k}^\alpha \qquad \text{(I)}$$

for any $\rho \in C^\infty(M,\mathbb{R})$; and

$$dx^\alpha_{i_1}\wedge\ldots\wedge dx^\alpha_{i_k} = (\mathrm{sgn}(\sigma))\, dx^\alpha_{\sigma(i_1)}\wedge\ldots\wedge dx^\alpha_{\sigma(i_k)}, \qquad \text{(II)}$$

where σ any permutation of $\{i_1,\ldots,i_k\}$ (and with the convention that 0-forms are smooth functions i.e. $\Lambda^0(M) := C^\infty(M,\mathbb{R})$).

The linear maps d_k: $\Lambda^k(M) \to \Lambda^{k+1}(M)$, which are defined on the spanning set by

$$d_k(\rho\, dx^\alpha_{i_1}\wedge\ldots\wedge dx^\alpha_{i_k}) = \sum_{i=1}^{d}\frac{\partial \rho}{\partial x_i} dx^\alpha_{i_m}\wedge dx^\alpha_{i_1}\wedge\ldots\wedge dx^\alpha_{i_k}$$

are called *exterior derivatives* and have the important property that $d_{k+1}\circ d_k \equiv 0$ (i.e. $\mathrm{image}(d_k)\subset\ker(d_{k+1})\subset\Lambda^{k+1}(M)$, as subspaces). This brings us to the main application.

Theorem (de Rham). The quotient space $H^k(M) = \ker(d_k)/\mathrm{image}(d_{k-1})$ is isomorphic to the kth (singular) cohomology group of M (over the reals).

Remark. Another important operation is the *star operation* $*:\Lambda^k(M)\longrightarrow\Lambda^{d-k}(M)$ defined on the bases by

$$*(dx^\alpha_{i_1}\wedge\ldots\wedge dx^\alpha_{i_k}) = \pm dx^\alpha_{j_1}\wedge\ldots\wedge dx^\alpha_{j_{d-k}},$$

where $\{j_1, \ldots, j_{d-k}\} = \{1, \ldots, d\} - \{i_1, \ldots, i_k\}$ and the sign is determined by the ordering.

A good reference for all of this material is [Warner].

B.3. Riemannian metrics and connections.

We define a C^∞ *Riemannian metric* $< , >_x$ to be an inner product associated to each tangent space T_xM (and with corresponding norm $\|v\|_x = <v,v>_x^{1/2}$, $v\in T_xM$) such that $x\mapsto <\frac{\partial}{\partial x^\alpha_i}, \frac{\partial}{\partial x^\alpha_j}>_x$ are C^∞ maps.

Theorem (Existence of Riemannian metrics). Any compact manifold M has a Riemannian metric.

(The proof of this fundamental result is rather easy: just patch together the natural inner product structure on co-ordinate patchs using a partition of unity.)

Length. Given a differentiable curve $c:[0,1]\longrightarrow M$ we can now define its length as

$$\ell(c)=\int_0^1\|c'(t)\|dt$$

(This definition being independent of the actual parameterisation of the curve.)

A closely related object is an *affine connection*, which is a map $(X,Y)\mapsto\nabla_X Y$, from $\mathfrak{X}^\infty(M)\times\mathfrak{X}^\infty(M)$ to $\mathfrak{X}^\infty(M)$, satisfying:

(i) $\nabla_{(fX+gY)}Z = f\nabla_X Z + g.\nabla_Y Z$;

(ii) $\nabla_X(Y+Z) = \nabla_X Y + \nabla_X Z$; and

(iii) $\nabla_X(fY) = f\nabla_X Y + X(f)Y$ where $\left(\sum_{i=1}^d u_i\frac{\partial}{\partial x_i^\alpha}\right)(f) = \sum_{i=1}^d u_i\frac{\partial f}{\partial x_i^\alpha}\in\mathbb{R}$,

for $X,Y,Z\in\mathfrak{X}^\infty(M)$ and $f,g\in C^\infty(M)$.

Definition. For a given basis $\left\{\frac{\partial}{\partial x_i^\alpha}\right\}_{i=1}^d$ we define the *Christoffel symbols* Γ_{ij}^k $(1\leq i,j,k\leq d)$ to be the coefficents in the expansions

$$\nabla_{\left\{\frac{\partial}{\partial x_i^\alpha}\right\}}\left(\frac{\partial}{\partial x_j^\alpha}\right)=\sum_{k=1}^d\Gamma_{ij}^k.\frac{\partial}{\partial x_k^\alpha}$$

The correspondence between Riemannian metrics and affine connections is given by the following fundamental result.

Theorem (Existence of affine connections). Given a Riemannian metric $<.,.>$ there is a *unique* affine connection such that:

(a) $\Gamma_{ij}^k=\Gamma_{ji}^k$, $\forall 1\leq i,j,k\leq d$ (Symmetry)

(b) $X<Y,Z> = <\nabla_X Y, Z> + <Y, \nabla_X Z>$, $\forall X,Y,Z\in\mathfrak{X}^\infty(M)$ (Levi-Civita identity)

(A very nice account appears in Chapter 3 of [Milnor].)

B.4 Geodesics and the exponential map.

Assume that M is a C^∞ compact manifold with a choice of local co-ordinates, and consider the solution curves $x(t)=(x_1(t),\cdots,x_d(t))$ to the differential equations

$$\frac{d^2 x_k}{dt^2} + \sum_{i,j=1}^{d} \Gamma_{ij}^k(x(t)) \frac{dx_i}{dt}\frac{dx_j}{dt} = 0, \quad k=1, \ldots, d$$

Using the local co-ordinates, these solutions correspond to curves $\gamma_v(t)\in M$, with $\gamma_v(0)=x$, $\gamma_v'(0) = v \in T_x M$, called *geodesics*.

For $x\in M$ and $\epsilon>0$, we let $B_\epsilon(x) = \{v\in T_x M \mid \|v\|_x \leq \epsilon\}$ denote the ϵ-ball in the fiber $T_x M$. We define the *exponential map* $\exp_x: B_\epsilon(x) \longrightarrow M$ by $\exp_x(v) = \gamma_v(1)$.

Theorem (Projecting from the tangent space to the manifold). If $\epsilon>0$ is sufficiently small then $\exp_x$: $B_\epsilon(x) \longrightarrow M$ is a C^∞ diffeomorphism onto its image.

Appendix C

Geodesic flows

In this final brief appendix we give a summary of the basic theory of geodesic flows. Geodesic flows are important examples which bridge ergodic theory and differential geometry. A general reference is [Anosov].

C.1 The geodesic flow.

We define a one-parameter family of diffeomorphisms ϕ_t: $TM \longrightarrow TM$ by $\phi_t(v) = \gamma_v{}'(t)$. These have the properties

(a) $\phi_s \circ \phi_t = \phi_{s+t}$ (i.e. the flow property), and

(b) ϕ_t preserves the *sphere bundle* $SM = \{ v \in TM \mid \|v\|_x = 1\}$

We define ϕ_t: $SM \longrightarrow SM$ to be the *geodesic flow* associated to M.

We can define a *volume* measure Vol_M on M locally, as follows. If $f \in C^0(M,\mathbb{R})$ with support $\mathrm{supp}(f) \subseteq U^\alpha$ then we define

$$\int f\, d(\mathrm{Vol}_M) = \int_{V^\alpha} f \circ (x^\alpha)^{-1}.\sqrt{\det(g_{ij})}\, dx_1 \ldots dx_d$$

where x^α: $U^\alpha \longrightarrow V^\alpha$ is a chart and (g_{ij}) is the matrix with entries $g_{ij} = < \frac{\partial}{\partial x_i^\alpha}, \frac{\partial}{\partial x_j^\alpha} >$, $1 \leq i,j \leq d$. (We then extend this definition to the whole manifold using a partition of unity.) If we now identify $S^{d-1} \equiv S_x M = \{v \in T_x M \mid \|v\| = 1\}$ then we can define the *Liouville measure* μ on SM by

$$\int F\, d\mu = \int_M \left\{ \int_{S_x M} f(x,v)\, d(\mathrm{Vol}_{S^{d-1}}) \right\} d(\mathrm{Vol}_M)(x)\ , \quad \forall F \in C^0(SM,\mathbb{R})$$

Theorem. The finite measure μ is invariant under the flow ϕ (i.e. $\int F \circ \phi_t d\mu = \int F d\mu$, $\forall F \in C^0(SM,\mathbb{R})$, $\forall t \in \mathbb{R}$).

We shall assume that that the measure μ is normalized so that $\mu(SM) = 1$.

C.2 Ergodicity and strict negative curvature.

Given two distinct (normalised) base vectors $\frac{\partial}{\partial x_i^\alpha}, \frac{\partial}{\partial x_j^\alpha} \in \mathfrak{X}^\infty(M)$ (in local co-ordinates) we can define the associated *sectional curvature* (up to normalization) to be the real-valued function on M given by

$$\kappa_{ij}(x) = <\left(\nabla_{\left\{\frac{\partial}{\partial x_j^\alpha}\right\}}\nabla_{\left\{\frac{\partial}{\partial x_i^\alpha}\right\}} - \nabla_{\left\{\frac{\partial}{\partial x_i^\alpha}\right\}}\nabla_{\left\{\frac{\partial}{\partial x_j^\alpha}\right\}}\right)\frac{\partial}{\partial x_i^\alpha}, \frac{\partial}{\partial x_j^\alpha}\Big)$$

We say that the sectional curvatures are negative if $\kappa_{ij}(x) < 0$, $\forall x \in M$, $\forall i \neq j$.

Theorem. If the sectional curvatures are negative then ϕ_t: $(SM,\mu) \longrightarrow (SM,\mu)$ is ergodic ($\forall\ t \neq 0$).

C.3 Ergodicity and non-positive curvature.

Consider the case where $\kappa_{ij}(x) \leq 0$; then there is the following well-known conjecture.

Conjecture A. The Liouville measure μ is ergodic if and only if the manifold V is *not* locally symmetric with rank greater than or equal to 2.

Pesin introduced the set $\Lambda = \{v \in M \mid v \textit{ has a non-zero Liapunov exponent for } g\}$, for which there is the following conjecture.

Conjecture B. $\mu(\Lambda) = 1$ if and only if the manifold V is *not* locally symmetric with rank greater than or equal to 2.

Ballmann, Brin and Eberlein have shown that Conjecture B implies Conjecture A [Ba-Br-Eb]).

Ballmann and Burns-Spatzier showed that manifolds of non-positive curvature *either* are locally symmetric manifolds of rank greater than or equal to two *or* satisfied a condition called *rank one* [Ballman],[Bur-Spa]. For example, any manifold with *strictly negative* sectional curvatures is automatically of rank one. However, rank one manifolds may also have large regions of zero curvature. Consider the following simple example: Take two flat tori T_1 and T_2 from each of which a disc has been removed. Next take a cylinder C with strictly negative curvature, except at its boundary where the curvature tends to zero. Let $M = T_1 \# C \# T_2$ be the surface formed by identifying the boundaries as illustrated in Figure 35. This is a rank one surface.

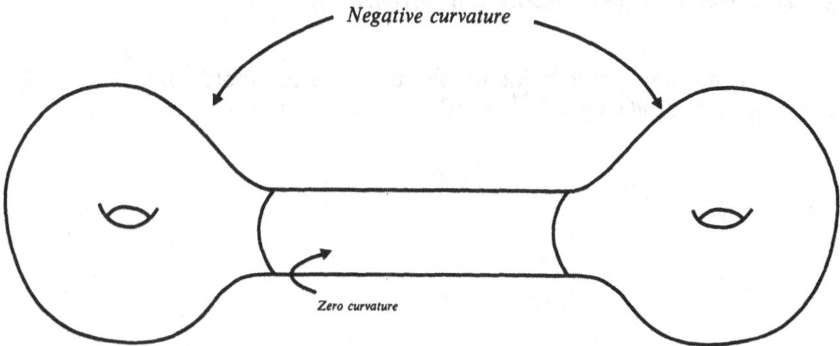

Figure 35: A rank one surface

C.4 Stable manifolds and non-positive curvature.

The special geometric nature of geodesic flows means that we can give slightly stronger results on their stable manifolds than those for arbitary flows. We say that vector $v \in M$ is *uniformly recurrent* if for each neighborhood $v \in U$ we have

$$\varliminf_{T \to +\infty} \frac{1}{T} \int_0^T \chi_U(g_t v)\, dt > 0$$

The following proposition tells us about the stable manifolds through uniformly recurrent vectors:

Proposition (Ballmann, Brin, and Eberlein). For any manifold M with non-positive curvature of rank one the uniformly recurrent vectors form a dense G_δ subset of M. Furthermore, for every uniformly recurrent vector $v \in M$ the stable manifold

$$W^s(v) = \{w \in M \mid d(g_t v,\, g_t w) \leq C\, d(v,w)\, e^{-\lambda t},\ t \geq 0\}$$

is a submanifold diffeomorphic to $\mathbf{R}^k$, where $C = C(v)$ and $\lambda = \lambda(v)$ depend on the choice of v (see [Ba-Br-Eb], p.186,192).

A similar result holds for the unstable manifolds $W^s(v) = \{w \in M \mid d(g_t v,\, g_t w) \leq C\, d(v,w)\, e^{-\lambda t},\ t \geq 0\}$.

References

[Anosov] V. Anosov, Geodesic flows on closed Riemannian manifolds with negative curvature, *Proc. Stelklov Institute*, 90 (1969) 1-235

[Ballmann] W. Ballmann, Non-positively curved manifolds of higher rank, *Ann. Math.*, 122 (1985) 597-609

[Ba-Br-Eb] W. Ballman, M. Brin, and P. Eberlein, Structure of manifolds of non-positive curvature, *Ann. Math.*, 122 (1985) 171-203

[Bed-Smi] E. Bedford and J. Smillie, Polynomial diffeomorphisms of C^2, *Invent. Math.*, 103 (1991) 69-99

[Ben-Car] M.Benedicks and L.Carleson, The dynamics of the Henon map, *Ann. Math.*, 133 (1991) 73-169

[Billingsley] P.Billingsley, *Ergodic Theory and Information*, Wiley, New York, 1965

[Bowen] R. Bowen, *Equilibrium States and the Ergodic Theory of Anosov Diffeomorphisms*, LNM 470, Springer, Berlin, 1975

[Bur-Spa] K. Burns and R. Spatzier, Manifolds of non-positive curvature and their buildings, *Pub. Math. (IHES)*, 65 (1987) 35-59

[Co-Fo-Si] I. Cornfeld, S. Fomin, and Y. Sinai, *Ergodic Theory*, Springer, Berlin, 1982

[Devaney] R. Devaney, *Introduction to Chaotic Dynamical Sytems*, Addison-Wesley, New York, 1987

[Fa-He-Yo] A. Fathi, M. Herman, and J-C Yoccoz, A proof of Pesin's stable manifold theorem, in LNM 1007, Springer, Berlin, 1983

[Fri-Mil] S. Friedland and J. Milnor, Dynamical properties of plane polynomial automorphisms, *Ergod. Th. Dynam. Sys.*, 9 (1989) 67-99

[Furstenberg] H. Furstenberg, Rigidity and cocycles for ergodic actions of semi-simple Lie groups, Seminaire Bourbaki No. 559, 1979-80

[Ga-Ha-La] S.Gallot, D.Hulin and J.Lafontaine, *Riemannian Geometry*, Springer, Berlin, 1987

[Gromov] M. Gromov, Entropy, Homology and Semi-algebraic geometry, Seminaire Bourbaki No. 663, 1985-1986

[Handel] M. Handel, The entropy of orientation preserving homeomorphisms, *Topology*, 21 (1982) 291-296

[Hopf] E. Hopf, *Ergodentheorie*, Chelsea, New York, 1937

[Irwin] M. Irwin, *Smooth Dynamical Systems*, Academic Press, London, 1980

[Ktn-Wei] Katznelson and B. Weiss, A simple proof of some ergodic theorems, *Israel J. Math.*, 42 (1982) 291-296

[Katok] A. Katok, Lyapunov exponents, entropy and periodic orbits for diffeomorphisms, *Pub. Math. (IHES)*, 51 (1980) 137-173

[Kat-Men] A. Katok and L. Mendoza, Smooth Ergodic Theory, unpublished notes

[Kat-Str] A. Katok and J-M Strelcyn, *Invariant Manifolds, Entropy and Billiards, Smooth Maps with Singularities*, LNM 1222, Springer, Berlin, 1988

[Led-You] F. Ledrappier and L-S. Young, The metric entropy of diffeomorphisms, *Ann. Math.*, 122 (1985) 509-574

[Mañé$_1$] R. Mañé, Lyapunov exponents and stable manifolds for compact transformations, in LNM 1007, Springer, Berlin, 1983

[Mañé$_2$] R. Mañé, *Teoria Ergodica*, IMPA, Rio de Janeiro, 1983

[Margulis] G. Margulis, *Discrete Subgroups of Semi-simple Lie Groups*, Springer, Berlin, 1991

[Mather] J. Mather, Characterization of Anosov diffeomorphisms, *Indag. Math.*, 30 (1968) 479-483

[Mendoza] L. Mendoza, Ergodic attractors for diffeomorphisms of surfaces, preprint

[Milnor] J. Milnor, *Morse Thoery*, Princeton University Press, 1963

[Misiurewicz] M. Misiurewicz, Diffeomorphisms without any measure of maximal entropy, *Bull. Acad. Polon. Sci. Ser. Math. Astron. Phys.*, 21 (1973) 903-910

[Newhouse_1] S. Newhouse, Continuity properties of entropy, *Ann. Math.*, 129 (1989) 215-237

[Newhouse_2] S. Newhouse, Entropy and Volume, *Ergodic Th. Dynam. Sys.*, 8 (1989) 283-299

[Newhouse_3] S. Newhouse, *Lectures in dynamical systems*, Progress in Math vol. 8, Birkhauser, Basel, 1980

[Nitecki] Z. Nitecki, *Differentiable Dynamics*, MIT Press, Cambridge, Mass., 1971

[Parry] W. Parry, *Topics in Ergodic Theory*, Cambridge University Press, 1981

[Parthasarathy] K. Parthasarathy, *Introduction to Probability and Measure*, Macmillan, London, 1980

[Pesin_1] Y. Pesin, Lyapunov characteristic exponents and ergodic properties of smooth dynamical systems with an invariant measure, *Sov. Math. Dok.*, 17 (1976) 196-199

[Pesin_2] Y. Pesin, Families of invariant manifolds corresponding to non-zero Liapunov exponents, *Izvestija*, 10 (1976) 1261-1305

[Pes-Sin] Y. Pesin and Y. Sinai, Hyperbolicity and stochasticity of dynamical systems, *Math. Phys. Rev.*, 2 (1981) 53-115

[Phelps] R. Phelps, *Lectures on Choquet's Theorem*, Van Nostrand, Princeton, 1966

[Pug-Shu] C. Pugh and M. Shub, Ergodic attractors, *Trans. Am. Math. Soc.*, 308 (1988)

[Rees] M. Rees, A minimal positive entropy homeomorphism of the 2-torus, *J. Lond. Math. Soc.*, 23 (1981) 537-550

[Rudin] W. Rudin, *Functional Analysis*, Tata-McGraw Hill, New Delhi, 1973

[Ruelle$_1$] D. Ruelle, Ergodic theory of differentiable dynamical systems, *Pub. Math.*, 50 (1981) 27-58

[Ruelle$_2$] D. Ruelle, An inequality for the entropy of differentiable maps, *Bol. Soc. Bras. Mat.*, 9 (1978) 83-87

[Ruelle$_3$] D. Ruelle, *Chaotic Evolution and Strange Attractors*, Cambridge University Press, 1989

[Ruelle$_4$] D. Ruelle, Characteristic exponents and invariant manifolds in Hilbert space, *Ann. of Math.*, 115 (1982)

[Shub] M. Shub, *Global Stability of Dynamical Systems*, Springer, Berlin, 1987

[Sinai$_1$] Y.Sinai, Construction of Markov partitions, *Fun. Anal. App.*, 2 (1968) 245-253

[Sinai$_2$] Y.Sinai, *Introduction to ergodic theory*, Princeton University Press, 1976

[Smale] S. Smale, Differentiable dynamics, *Bull. Am. Math. Soc.*, 73 (1967) 97-116

[Walters] P. Walters, *An Introduction to Ergodic Theory*, GTM 79, Springer, Berlin, 1982

[Warner] Warner, *Foundations of Differentiable Manifolds and Lie Groups*, Springer, Berlin, 1980

[Whitley] D. Whitley, Discrete dynamical systems in dimensions one and two, *Bull. Lond. Math. Soc.*, 15 (1983) 177-217

[Yomdin] Y. Yomdin, Volume growth and entropy, *Israel J. Math.*, 57 (1987) 285-318

[Zimmer] R. Zimmer, *Ergodic Theory and Semi-simple Groups*, Birhauser, Berlin, 1985

Index

absolute continuity 131
-Pesin's theorem 132
affine connections 149
Alaoglu's theorem 143
algebra 139
axiom Λ diffeomorphisms 84
-examples 84
Birkhoff ergodic theorem 12
-proof 13–18
Christoffel symbols 149
closing lemma (hyperbolic) 84
closing lemma (non-hyperbolic) 95
-proof 95
conditional expectation 18
continued fractions 14
converse Birkhoff theorem 129
de Rham cohomology 148
Dirac measure 12, 140
entropy rigidity (for Lie groups) 61
entropy stability 108
-Katok's theorem 109
-Yomdin's theorem 109
ergodic components 132
-examples 132–133
-absolute continuity 133
ergodic measures 9
-characterising 12
-existance 8
exponential map 150
exterior derivative 148
forms 147
Gauss map 7
Gauss measure 7
geodesic 150
geodesic flow 151
global stable manifolds 124
-examples 124–126
Henon map 79–81
Holder continuity 71
homoclinic point 103
-existance 105
-location 105
homology 52
hyperbolic fixed points 103
hyperbolic measures 63
invariant measures 5
-characterising 6
-existance 8
Ledrappier—Young theorem 128
-example 128–129
Liapunov exponents (for surfaces) 23
Liapunov exponents (arbitrary dimension) 23
-examples 25–31
Liapunov metric 70
Liapunov neighbourhood 88
Lie groups 60
Liouville measure 151
local co-ordinates (charts) 145
-bases 147
local product structure 126
manifold 145
measurable maps 141
measurable spaces 5, 140
measurable transformations 141
measure 140
-absolutely continuous 142
-equivalent 142
-mutually singular 142
-support 143
measure theoretic entropy (Katok's definition) 43–44
-Kolmogorov—Sinai definition 53
-equivalence of definitions 54–57
-examples 44–45
-history 45–46
measurable space 140
non-uniformly hyperbolic diffeomorphisms 28
normal numbers 14
Osceledic theorem (for surfaces) 23

–proof 31–36
Osceledic theorem (arbitrary dimension) 24
–generalizations 36
periodic points 96
–existance 96–97
–examples 97
–location 100
–proof 119
–examples 100–101
–asymptotics 101
–sketch proof 120
–examples 102–103
Pesin set 64
–examples 65–68
Pesin–Ruelle inequality 46
–proof 58–60
–equality 47, 127–128
–strict inequality 47–48
Pesin's theorem 127
–example 127
Plykin attractor 30
Poincare Recurrence 9
polynomial diffeomorphisms 81
pseudo–orbit (hyperbolic) 84
pseudo–orbit (non-hyperbolic) 90–91
Riemannian metric 148
–existance 149
–length of curves 149
sectional curvature 152
shadowing lemma (hyperbolic) 84
shadowing lemma (non-hyperbolic) 91
–proof 92–93
shadowing point 91
–existance 91
–uniqueness 94
Shannon's theorem 54
Shub entropy conjecture 111
–Yomdin's solution 112
–sketch proof 113–114
–continuous counter-example 112
sigma algebra 140
Smale homoclinic point 107
Smale horse-shoe 27
–generalized 105
–existance 108
–sketch proof 120–121
–non-uniformly hyperbolic 28–29
sphere bundle 151
stable manifolds 85
–for fixed points 103
–for Pesin sets 123
stable manifold theorem 124
–proof 133–136
star operation 148
subadditive ergodic theorem 22
–proof 36–40
subadditivity 30
tangent bundle 147
topological entropy 48–49
–examples 49
–history 49–50
tracing point 84
transverse intersection 89
triangulation of manifolds 146
uniformly hyperbolic diffeomorpisms 28, 84
uniformly recurrent vectors 152
upper semi-continuity of topological entropy 115
–examples 116–119
variational principle 52
vector bundles 146
–fibres 146
vector fields 147
volume 151
weak star topology 143
Whitney embedding theorem 146
Whitney sum 146
Yomdin's inequality 110

Printed in the United States
By Bookmasters